T0135934

Processing of relevant characteristics of complex sounds in normal-hearing listeners and cochlear implant users

Dissertation

zur Erlangung des akademischen Grades

doctor rerum naturalium
(Dr. rer. nat.)

genehmigt durch
die Fakultät für Naturwissenschaften
der Otto-von-Guericke-Universität Magdeburg

von: Dipl.-Ing. (FH) Wiebke Heeren, M.Sc.

geb. am: 09. September 1984 in Wilhelmshaven

Gutachter: Prof. Dr. rer. nat. Jesko L. Verhey

                Prof. Dr. rer. nat. Sebastian Hoth

eingereicht am: 26. August 2014

verteidigt am: 30. März 2015

Bibliographic information published by the Deutsche Nationalbibliothek

The Deutsche Nationalbibliothek lists this publication in the Deutsche Nationalbibliografie; detailed bibliographic data are available on the Internet at http://dnb.d-nb.de .

ISBN 978-3-8325-3997-9

Logos Verlag Berlin GmbH
Comeniushof, Gubener Str. 47,
10243 Berlin
Tel.: +49 (0)30 42 85 10 90
Fax: +49 (0)30 42 85 10 92
INTERNET: http://www.logos-verlag.de

Kurzfassung

Die vorliegende Arbeit befasst sich mit der Verarbeitung verschiedener Charakteristika von komplexen Schallen bei Normalhörenden und Cochlea-Implantat-Trägern (CI-Träger). Trotz jahrzehntelanger Forschung ist die auditorische Verarbeitung komplexer Schalle noch nicht vollständig verstanden. Daher ist eine klarere Einsicht darin, wie das auditorische System Schalle verarbeitet sowie ein besseres Verständnis der Zusammenhänge zwischen physikalischen Parametern und dem entstehenden Höreindruck noch immer von großem Interesse. Viele Studien beschäftigen sich mit der Wahrnehmung unterschiedlicher Geräusche an der Schwelle. Andere hingegen erforschen ob und wie sich die Ergebnisse dieser schwellennahen Untersuchungen auf die überschwellige Wahrnehmung übertragen lassen. Solche Zusammenhänge lassen sich beispielsweise mit Hilfe psychoakustischer Experimente quantifizieren. Der Fokus der vorliegenden Arbeit liegt in der Beschreibung der überschwelligen Wahrnehmung von Schallereignissen mit unterschiedlichen spektralen, zeitlichen und räumlichen Eigenschaften mittels ebensolcher psychoakustischer Experimente. Dabei wird sowohl die Wahrnehmung im normalen Gehör als auch bei CI-Trägern näher betrachtet. Cochlea-Implantate (CIs) bieten die einzige Möglichkeit das Hörvermögen von Menschen mit hochgradigem bis vollständigem Hörverlust zumindest teilweise wieder herzustellen. Trotz großer Fortschritte beispielsweise im Sprachverständnis, ist weitere Forschung z.B. in Bezug auf räumliche Wahrnehmung und Quellentrennung erforderlich.

Eine mögliche Beschreibung überschwelliger Wahrnehmung im Hinblick auf spektrale, zeitliche und räumliche Eigenschaften von Schallsignalen ist die Beschreibung der Lautheit solcher Signale, da diese nicht nur von der Intensität sondern auch von spektralen und zeitlichen Aspekten abhängig ist. Mit zunehmender Bandbreite eines Signals bei gleichbleibender Energie steigt beispielsweise auch die wahrgenommene Lautheit an. Dieser als spektrale Lautheitssummation bezeichnete Effekt wird für gewöhnlich für Schalle gemessen, bei denen alle Frequenzkomponenten gleichzeitig dargeboten werden. In der vorliegenden Arbeit wird unter anderem gezeigt, dass dieser Effekt auch für Schalle nachweisbar ist, deren Frequenzkomponenten zu unterschiedlichen Zeitpunkten dargeboten werden. Die Ergebnisse unterstützen die These von Zwicker (1969), dass einzelne Tonpulse eine schnell ansteigende spezifische Lautheit hervorrufen, die nur langsam wieder abklingt, wodurch sich die Lautheiten auch bei zeitversetzter Darbietung aufaddieren. Neben spektralen Effekten befasst sich die vorliegende Arbeit auch mit zeitlichen Aspekten der Lautheit. Neben der Integration über verschiedene Frequenzen integriert das auditorische System auch über die Dauer eines Signals: Der Pegel eines kurzen Signals muss höher sein als der eines langen Signales um die gleiche Lautheit hervorzurufen (Port, 1963a). Des Weiteren wird die Gesamtlautheit eines Signals auch vom Pegel oder spektralen Gehalt einzelner zeitlicher Segmente bestimmt. Bisherige Studien haben diese zeitlichen und spektralen Wichtungen nur separat betrachtet. In der vorliegenden Arbeit hingegen werden mittels zeitlicher Pegelvariation in verschiedenen spektralen Regionen spektro-temporale Wichtungen ermittelt die zur globalen Lautheitsbewertung beitragen. Die Ergebnisse zeigen, dass die spektro-temporalen Wichtungen durch getrennt ermittelte spektrale und zeitliche Wichtungen vorhergesagt werden können. Dies lässt darauf schließen, dass die zeitlichen und spektralen Wichtungen komplexer Schalle von einander unabhängige Prozesse darstellen.

Eine Messmethodik zur Bestimmung der Lautheit von Signalen ist die Kategoriale Lautheitsskalierung (KLS, Brand and Hohmann, 2002), mit Hilfe derer die Lautheitsempfindung über den kompletten auditorischen Dynamikbereich bestimmt werden kann. Durch Unterschiede in den resultierenden Lautheitsfunktionen kann beispielsweise auch eine Differenzierung zwischen normalem und pathologischem Gehör erfolgen. Allerdings werden die Ergebnisse der Skalierungen in kategorialen Einheiten (categorical units, CU) angegeben. Die führt dazu, dass ein direkter Vergleich mit den Ergebnissen aktueller Lautheitsmodelle, deren Ergebnisse in der Einheit Sone angegeben werden, nicht ohne Weiteres möglich ist. Kapitel 4 dieser Arbeit versucht die verschiedenen Einheiten zueinander in Beziehung zu setzen und stellt eine Umrechnungsformel vor, welche die Übertragung von Sone auf die CU-Lautheit ermöglicht.

Während für die zuvor beschriebenen Messungen Reintöne oder Schmalbandrauschen verwendet werden, sind wir im Alltag umgeben von komplexen akustischen Situationen, in denen mehrere Schallereignisse gleichzeitig auftreten. Zumeist interessiert dabei nur die Information eines einziges Signals wie beispielsweise Sprache in lauter Umgebung oder ein Sprecher unter vielen. Der Einfluss unterschiedlicher Störgeräusche auf die Signalempfindung wird oftmals mit Hilfe von durch Rauschen maskierten Sinustönen erforscht. Dabei können zeitliche Fluktuationen des Maskierers oder die räumliche Trennung von Signal und Maskierer zu einem Vorteil in der Signaldetektion führen (release from masking). Diese Effekte werden zumeist nur für die Maskierungsschwelle des Signals betrachtet. Mit Hilfe der kategorialen Lautheitsskalierung wird in der vorliegenden Arbeit die Auswirkung dieser Effekte auf die überschwellige Wahrnehmung sowohl bei Normalhörenden als auch für CI-Träger untersucht. Die Ergebnisse zwei der Studien dieser Arbeit zeigen eine reduzierte Maskierwirkung bei Normalhörenden für überschwellige Pegel bis zu 30 dB. Ein solcher binauraler Gewinn kann bei den beidseitig versorgten CI-Trägern jedoch nicht beobachtet werden. Dieses Ergebnis zeigt zwar noch bestehende Probleme bei der binauralen Verarbeitung in Cochlea-Implantaten auf, dennoch kann in der vorliegenden Arbeit auch gezeigt werden, dass ein gutes räumliches Hören mit

CIs möglich ist. In der abschließenden Studie dieser Arbeit wird mittels eines Lautsprecherrings bestehend aus 31 im Halbkreis angeordneten Lautsprechern und virtueller Signalgenerierung das räumliche Auflösungsvermögen von CI-Trägern und Normalhörenden für statische und bewegte Quellen untersucht. Im Vergleich zur normalhörenden Kontrollgruppe zeigen die CI-Träger zwar ein etwas schlechteres räumliches Auflösungsvermögen, dennoch zeigen beide Gruppen vergleichbare Tendenzen in der Abhängigkeit von Faktoren wie Winkeldifferenz, überstrichener Winkel oder Winkelgeschwindigkeit.

Abstract

The suprathreshold perception of acoustic stimuli depends, for instance, on the sounds' spectral, temporal and spatial aspects corresponding to the present soundfield. The present thesis investigates the suprathreshold perception of sounds with different spectral, temporal and spatial content in normal-hearing listeners and cochlear implant (CI) users.

The loudness of a sound depends for example on its duration and its spectral content. Effects describing these dependecies are referred to as temporal integration and spectral summation of loudness. In the first two studies of this thesis it is shown that (i) spectral loudness summation also occurs for non-simultaneously presented tone pulses of different spectral content and (ii) that loudness also depends on the level or spectral content in different temporal segments and that both dependencies are processed independently.

A measurement approach that is suitable for investigations on suprathreshold perception is the categorical loudness scaling (CLS) procedure. However, resulting loudness functions are given in terms of categorical units (CU), whereas, the loudness in current loudness models is given in sone. To enable a better comparison of corresponding data, a function is proposed linking loudness in CU to the classical measure sone. In addition, the CLS procedure is used to investigate the release from masking due to temporal envelope fluctuations and spatial separation of masking signals at signal levels above threshold. The results reveal that the effects observed at threshold also occur at suprathreshold levels up to about 30 dB above the threshold in a corresponding baseline condition. In a similar experiment with spatial separation of signal and masker in CI users, no differences in perception of the signal were found for the baseline and the masking release condition. In fact, it was observed that spectral similarity of signal and masker, which leads to a greater benefit in normal-hearing listeners, leads to problems with the perceptual separation of signal and masker in the CI users. Although these results reveal difficulties in binaural processing in CI users, the investigations in the last chapter indicate good performance for spatial discrimination of stationary and moving sound sources in CI users. Despite poorer performance as compared to normal-hearing listeners, similar trends and dependencies on sound properties are found in both groups of listeners.

Contents

Contents

1 Introduction

1.1 Loudness

Loudness is a subjective measure which describes the relation between a sound event and a reference sound and belongs to the category of intensity sensations. The unit used to measure loudness is sone. A sound has the loudness of 1 sone if it is perceived as loud as a 1-kHz sinusoid with a level of 40 dB SPL. A doubling of the loudness is achieved by a level increase of 10 dB SPL, i.e. a tenfold increase of the intensity. Thus, a 1-kHz sinusoid with a level of 50 dB SPL has a loudness of 2 sone. This simple relationship only holds for medium and high sound levels; at low levels a doubling in loudness is caused by a lower level increase. Figure 1.1 shows the loudness function for a 1-kHz sinusoid.

The relation between the objective sound intensity and the individually perceived loudness is described by a loudness function. According to Zwicker and Fastl (1999), this function can be estimated by a power law as follows:

$$N = \frac{1}{16} \left(\frac{I}{I_0} \right)^{0.3} \tag{1.1}$$

The exponent of 0.3 defines the compression of the loudness function. On a logarithmic scale, a doubling equals to an increase of 3 dB. The doubling of loudness is caused by an increase of sound level by 10 dB SPL. This leads to a relation in level of 3/10 which just corresponds to the exponent of the power law.

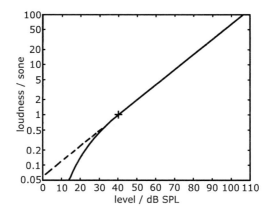

Figure 1.1: Loudness function: relation between loudness in sone and the level of a 1-kHz sinusoid (solid black line) and the estimation of the loudness function by the power law of Eq. 1.1; replotted from Zwicker and Fastl (1999).

Besides sone and the loudness function, the loudness of a sound can be described by equal loudness levels given in phon. The loudness level of a 1-kHz sinusoid is equal to its sound pressure level in dB SPL. The loudness level of a sound equals the sound pressure level of a 1-kHz sinusoid that is equally loud as the sound. The curves that connect equal loudness levels over different frequencies are referred to as equal-loudness-level contours (ELLC). The course of the ELLCs is plotted in Figure 1.2 for loudness levels between 10 and 100 phon. According to their definition, all ELLCs have to go through the sound pressure level at 1 kHz that equals the value of the parameter of the curve in phon. That implies that, e.g. the 40-phon ELLC has to go through 40 dB at 1 kHz. The threshold in quiet has also the shape of an ELLC. As for 1 kHz it amounts to 3 dB on average, the threshold in quiet equals the 3-phon ELLC.

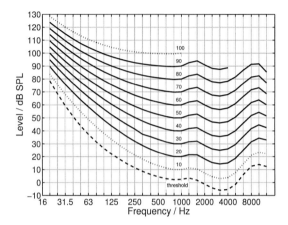

Figure 1.2: Equal-loudness level contours for loudness levels between 10 and 100 phon. The lowest curve equals the 3-phon ELLC and thus corresponds to the hearing threshold; replotted from ISO 226 (2003).

The ELLCs can also be related to the loudness in sone. As a 1-kHz sinusoid with a level of 40 dB SPL has a loudness of 1 sone, this also holds for the 40-phon ELLC. Accordingly, the 60-phon ELLC has a loudness of 4 sone, as an increase in level by 20 dB results in a four times higher loudness.

In addition to sone and phon loudness can be expressed in terms of categorical units (CU). Using a procedure named categorical loudness scaling (CLS), the loudness of a stimulus over the whole auditory dynamic range can be measured by using a scale with named categories like "inaudible", "very soft", "soft", "medium" etc., respectively. Originally, this procedure was invented by Heller (1985) who used a two step procedure starting with a rough followed by a more precise scale. A simplified version of this procedure using only one scale with more than 50 response alternatives was proposed by Hellbrück and Moser (1985). Another alternative scaling procedure using 11 response alternatives was invented by Hohmann and Kollmeier (1995) and improved with regard to measurement time and bias effects by Brand and Hohmann (2002). The latter one also serves as a basis for the international standard for categorical loudness scaling: ISO 16832 (2006). Basically, this scaling procedure consists of seven named categories from "inaudible" to "extremely loud", and four unnamed intermediate categories between "soft" and "loud". For further data analysis, these categories were linearly transformed into numerical values from 0 CU ("inaudible") to 50 CU ("extremely loud") in 5-CU steps. Subsequently, a function consisting of two linear parts and a smoothed transition region is fitted to the measurement values, resulting in a loudness (growths) function.

A relation between the different loudness measures stated above is given in Chapter 4.

1.1.1 Spectral aspects

The loudness evoked by a sound is not only influenced by its level, but also by its spectral content. The term *spectral loudness summation* describes the effect that the loudness of a sound with constant intensity increases with increasing bandwith. Earlier studies of Zwicker and Feldtkeller (1955), Zwicker *et al.* (1957), Fletcher and Munson (1933) and Verhey and Kollmeier (2002) showed that for equal loudness the level of a narrowband sound has to be higher than the level of a broadband sound. Increasing the bandwidth of a narrowband signal with constant energy content, the loudness remains constant up to a *critical* bandwidth. Above that bandwith, the loudness increases with increasing bandwidth. This can be described by the assumption that the intensity is devided into different frequency bands by overlapping bandpass filters with different center frequencies. By means of a non-linear, compressive function the intensity is transformed into *specific* loudness (i.e. loudness per frequency band) and subsequently added up across the frequency bands. This effect has already been implemented in loudness models for stationary sounds (e.g. Fletcher and Munson, 1937; Zwicker and Scharf, 1965; Moore and Glasberg, 1996). The general structure of such a loudness model is shown in Figure 1.3.

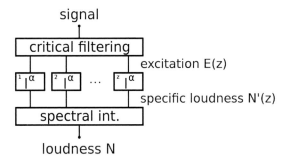

Figure 1.3: General structure of a stationary loudness model. It realizes the assumption of frequency-band filters with different critical bandwidth and the compressive behaviour and can thus model spectral loudness summation. Filtering the long-term spectrum frequency-band wise results in the excitation $E(z)$ per frequency band z. Processed by the compressive nonlinearity this results in the specific loudness $N'(z)$. The overall loudness N is then given by the subsequent spectral integration across the frequency bands; from Hots (2014).

As the loudness is compressive, a sound with an energy content located in one frequency band elicits a lower loudness than a sound with an energy content that is split up over several frequency bands. Generally, it holds that both sounds consist of the same number of spectral components I_i with the overall loudness $I = \sum_i I_i$. For bandwidths smaller than the corresponding frequency-band width the loudness is calculated according to

$$N_{narrow} = I^{0.3}. \qquad (1.2)$$

In case of a sound with a bandwidth spanning several frequency bands, the specific loudness of each band has to be considered:

$$N_{broad} = \sum_i I_i^{0.3}. \qquad (1.3)$$

Thus, the amount of the spectral loudness summation is defined by the width of the frequency bands and the depth of the compression:

$$N_{broad} = \sum_i I_i^{0.3} > \left(\sum_i I_i \right)^{0.3} = N_{narrow}. \qquad (1.4)$$

The width of the frequency bands is frequency dependend and increases with increasing frequency. The relation between center frequency and bandwidth can be described by ERB-scaling (equivalent rectangular bandwidth, according to Moore, 2003) and Bark-scaling (Zwicker and Fastl, 1999), respectively. Both scales merely differ in the width and thus in the number of frequency bands that cover the audible frequency range.

1.1.2 Temporal aspects

Apart from the spectral content, the loudness of a sound also depends on its duration. Port (1963a), Poulsen (1981), Florentine *et al.* (1996) and Buus *et al.* (1997) showed that for equal spectral content and equal level a long sound elicits a higher loudness than a short sound. This effect is referred to as temporal integration of loudness.

Port (1963a) investigated the loudness of noise stimuli with different bandwidths depending on the stimulus duration. For constant stimulus level, the results reveal an increasing loudness up to a duration of 70 ms. For longer durations the loudness remained constant. According to Port one reason for this findings might be that durations lower than 70 ms depend on the energy of the sound, i.e. the integration of the sound intensity over time. Whereas the loudness for longer durations depends on the sound intensity. Further measurements by Port indicated an independency of this time constant of 70 ms from spectral content and level of the stimulus. This was confirmed by measurements of Buus *et al.* (1997). Niese (1959) and Munson (1947) also obtained certain time constants for temporal loudness summation. However, they measured durations of 30 and 400 ms, respectively. Port attributed these differences to differences in the measurement procedures. Poulsen (1981) compared the loudness of stimuli with short durations with twice as long stimuli for levels up to 95 dB SPL. The results lead to time constants of about 100 ms close to threshold and 200 ms for levels clearly above threshold. However, using one time constant was not sufficient to explain the measurement data properly, especially for short stimulus durations. Thus, he suggested a model using a combination of both a short and a long time constant.

These results lead to an extension of the loudness model concept described above (Figure 1.3). Such a dynamic loudness model additionally comprises a temporal integration stage that integrates the loudness using a lowpass filter with a specific time constant. The general structure of such a model is shown in Figure 1.4. Preceding the spectral stage, the time signal of the sound is temporally windowed using time segments of

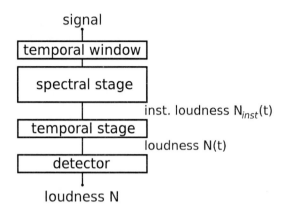

Figure 1.4: General structure of a dynamic loudness model. Previous to the spectral stage (see also Fig. 1.3) the time signal of the sound is temporally windowed leading to instantaneous loudness values $N_{inst}(t)$ for each time segment t. In the last step the overall loudness N of the sound is calculated by, e.g. determining the maximum or a specific percentile of the time dependent signal; from Hots (2014)

a few milliseconds. Each time segment is then processed in the spectral stage according to the stationary model described in section 1.1.1. This leads to instantaneous loudness values $N_{inst}(t)$ for each time segment t. In the following temporal integration stage these instantaneous loudness values are lowpass filtered, resulting in the loudness values $N(t)$. At last, the overall loudness N of the sound is calculated by, e.g. determining the maximum or a specific percentile of the time dependent signal. Different loudness models using this concept have been introduced (e.g. Zwicker, 1969; Ogura *et al.*, 1991; Glasberg and Moore, 2002; Chalupper, 2002). Zwicker realized the temporal integration for tone pulses by using a low-pass filter with a time constant of 100 ms. Ogura *et al.* picked up on this approach and used this 100-ms time-constant to describe the increase in loudness, and an additional time constant of 5 ms to describe the decay. This combination allowed for the prediction of the increase in loudness of noise-pulse trains with increasing repetition rate. In the temporal integration stage of the model for time-varying sounds (TVL) by Glasberg and Moore (2002), so-called short and long-term loudness were introduced. This enabled the prediction of the temporal integration of tone pulses and amplitude modulated signals (Rennies *et al.*, 2009).

1.1.3 Spectro-temporal aspects

Both, spectral and temporal effects on loudness summation do not occur separately. Combining both effects leads to spectro-temporal loudness summation. In their studies, Verhey and Kollmeier (2002), Anweiler and Verhey (2006) and Verhey and Uhlemann (2008) showed that the spectral loudness summation, i.e. the difference in level between broad and narrow band sounds at equal loudness, depends on the duration of the sounds with a lower amount for short sounds.

Verhey and Kollmeier (2002) investigated the spectral loudness summation of noise stimuli with different durations and bandwidhs at different reference levels. They compared equal loudness levels of stimuli with durations of 10, 100 and 1000 ms for bandwidths between 200 and 6400 Hz with a reference bandwidth of 3200 Hz. For all reference levels they found a higher loudness summation between the reference signal and the 200-Hz wide test signal for a duration of 10 ms compared to a duration of 1000 ms. To investigate the influence of the measurement procedure, Anweiler and Verhey (2006) measured the spectro-temporal loudness summation using categorical loudness scaling (CLS) and a loudness matching procedure. By means of

CLS, loudness functions of band pass noises with different bandwidths and durations of 10 and 1000 ms were determined. The results for reference levels between 25 and 105 dB SPL showed that the level-dependent relation in loudness between long and short stimuli differs for different spectral content. Similar results could be obtained using the loudness matching procedure.

On the basis of Verhey and Kollmeier (2002), Verhey and Uhlemann (2008) conducted investigations on spectral loudness summation of noise pulse trains with different bandwidths. They measured the level difference between equally loud noise pulse trains of different bandwidths to ascertain whether (a) the spectral loudness summation is equal to single tone pulses and whether (b) spectral loudness summation depends on the repetition rate. Therefore, they used a pulse duration of 10 ms, repetition rates between 3 and 100 Hz and bandwidths between 200 and 6400 Hz. In a preliminary experiment, single pulses with durations of 10 and 100 ms instead of pulse trains were used. The results of this experiment reveal evident spectral loudness summation for bandwidths greater than the reference bandwidth of 400 Hz. Furthermore, the level difference between the 200 Hz and the equally loud 6400-Hz wide noise pulse for a pulse duration of 10 ms is by 8 dB higher than for a pulse duration of 1000 ms. This is in accordance with the greater loudness summation for shorter signals described in Verhey and Kollmeier (2002). The measurements with different repetition rates revealed that for low repetition rates up to 50 Hz spectral loudness summation is approximately the same as for short single tone pulses. Above 50 Hz, spectral loudness summation decreased with increasing repetition rate. For a repetition rate of 100 Hz results are similar to those of long single tone pulses.

The studies described so far merely cover signals with frequency components presented simultaneously. On the contrary, Zwicker (1969) examined the influence of the temporal structure of tones and tone pulse trains with varying frequencies on the loudness. Among others, he compared the level of a 100 ms long tone pulse with a sequence of five 20 ms long pulses with frequencies of 1370, 2500, 1000, 3400 and 1850 Hz at equal loudness. To avoid systematic errors he did not present the frequencies in ascending but in the stated order. The results showed a distinct spectral loudness summation of about 11 dB. However, this level difference was smaller than the level difference of 23 dB at equal loudness between a 100 ms long single tone and a 100 ms long complex sound including all five stated frequencies. About 7 dB (=10lg5) of this larger difference can be explained by the increase in level due to five tones presented simultaneously compared to one single tone. Zwicker explained the lower difference for the tone pulse trains by the effect that the single pulses elicit a rapidly increasing loudness which slowly decreases. If the loudness of one pulse is not entirely decayed when the next pulse started, the partial loudness of the different frequency bands add up to a higher overall loudness. To verify this hypothesis of slowly decaying specific loudness, Rennies (2008) performed further measurements on forward masking and spectral loudness summation. As contrastet with Zwicker (1969), Rennies (2008) did not use continuous reference signals but signals with the same temporal structure as the test stimulus. The author varied the duration of the pause between the single pulses within one stimulus (inter-pulse interval, IPI) to test whether from a certain duration onwards the spectral loudness summation decreases. Therefore he used two different test conditions: condition 1 was comprised of five time-displaced tone pulses with a frequency of 3400, 1850, 2500, 1370 and 4800 Hz, respectively. Condition 2 had the same temporal structure, but each single pulse contained all five frequency components. Additionally, Rennies used tone pulses of equal loudness instead of equal level. The use of equal level could lead to a loudness of the pulse train which is dominated by the loudest frequency component (Rennies, 2008). The results, however, did not reveal an influence of the IPI on the level difference. To approve the data of Zwicker (1969) the level difference in condition 1 had to decrease with increasing IPI until the specific loudness of the preceeding pulse has entirely decayed when the next pulse started. For IPIs equal to and greater than this marginal IPI the level of the test and the reference stimulus should be equal for equal loudness. According to Rennies (2008), the differences in stimulus duration for different IPI could be one possible reason for the mismatch between the hypothesis and the results. Due to the increasing IPI, the longest stimulus duration was five times longer than the shortest. For further studies he suggested to use equally long stimuli and longer IPIs up to 190 ms. Amongst others, these suggestions are verified in Chapter 2 of this thesis.

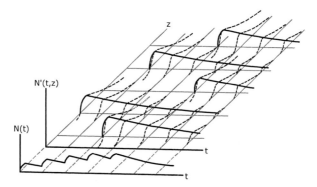

Figure 1.5: Schematic diagram of specific loudness $N'(f, t)$ as a function of time and frequency for a sequence of short tone pulses with different frequencies. In the background, the loudness $N(t)$ is indicated as the sum of $N'(f, t)$ across frequency (replotted from Zwicker (1969), Fig. 8)

In addition to their experiments, Verhey and Kollmeier (2002) and Rennies (2008) put the dynamic loudness models on the proof for spectro-temporal effects in loudness summation. Verhey and Kollmeier stated that in models which process the temporal integration separately from the spectral integration only at the end of the model, the spectral loudness summation remains independent of the duration of the signal. Even the dynamic loudness model (DLM) of Chalupper (2002) which applies one part of the temporal processing (forward masking) previous to the spectral stage cannot be taken into account for the duration dependence of the spectral loudness summation. According to Verhey and Kollmeier (2002) this is due to the fact that only the decay time of the forward-masking slope varies with duration which is not crutial for the spectral loudness summation of short noise pulses. However, according to Rennies (2008) the DLM should be preferred to the TVL for expansions and variations to simulate dynamic aspects of the spectral loudness summation as the temporal process depends at least partially on the spectrum. Thus, Rennies *et al.* (2009) introduced a bandwidth-dependent amplification prior to the spectral stage. With this extended dynamic loudness model (eDLM) they were able to show that it is possible to predict an increasing spectral loudness summation for short signals.

The perception of loudness can also be influenced by the amplitude modulation of a sound. Grimm *et al.* (2002) investigated the loudness of amplitude modulated sounds. For this purpose, they imprinted either sinusoidal or stochastical modulations on sinusoids and noises with different bandwidths and determined the dependency of the loudness on stimulus bandwidth and modulation frequency. For both types of amplitude modulation, a dependency of the loudness on modulation frequency reveals for frequencies between 4 and 32 Hz. For equal loudness, the modulated signal needed a slightly lower level. For the investigation of the bandwidth dependency Grimm *et al.*

used a fixed sinusoidal modulation with a frequency of 8 Hz and a fixed stochastical modulation with a frequency of 16 Hz. Both modulation types caused a decrease in the level difference between the unmodulated test signal and the modulated reference signal for increasing bandwidth. For narrowband signals the modulated signal had a slightly higher level than the unmodulated signal with equal bandwidth at equal loudness. On the contrary, for broadband signals the modulated signal had to be slightly lower in level to evoke equal loudness. Thereby, stochastical modulation has higher impact on narrowband signals, whereas for broadband signals the impact of sinusoidal modulation is higher. Grimm *et al.* explained their results with the assumption that modulated signals could be taken as a sequence of short signals. This would evoke a higher loudness than an (unmodulated) continuous signal.

Further spectro-temporal effects on loudness, especially of non-simultaneous tone pulses and spectro-temporal weighting, are discussed in Chaps. 2 and 3.

1.2 Spatial hearing

The ability to localize sounds is of great importance to determine the direction of objects and thus to be able to draw visual attention to them. In general, there are two different aspects that describe the performance in localization. On the one hand, the difference between the perceived and the actual direction of the sound, thus measuring the absolute error of the auditory sytem, and on the other hand, the smallest detectable change in angular position or minimum audible angle (MAA, in this thesis also referred to as static angle resolution) that describes the resolution of the auditory system (Mills, 1958). The localization ability depends on different characteristics of the sound such as time and level differences between the two ears or the spectral characteristics of the sound and/ or the background noise. As most of the surrounding sounds are generally not emitted from stationary but from moving sound sources, an additional attribute to be considered is the motion of the sound source. This can be verified by determining the smallest change in location of a moving stimulus (minimum audible movement angle, MAMA, in this thesis also referred to as dynamic angle resolution).

1.2.1 Localization cues

Two main cues for the localization of sound sources are the interaural level differences (ILD) and the interaural time differences (ITD). The former mainly occurs at high frequencies, since for very low frequencies (especially below 500 Hz) the wavelength of the sound is long compared to the dimensions of the head. Thus there is little or no influence on the amplitude of the sound. For higher frequencies, the wavelength of the sound becomes shorter compared to the dimensions of the head leading to shadow effects cast by the head due to diffraction. Feddersen *et al.* (1957) determined level differences up to 20 dB at at high frequencies around 6 kHz for a direction of sound incidence of ±90°. In other studies these maximum level differences were considerably greater. In Middlebrooks *et al.* (1989), e.g., maximum ILDs measured near 90° azimuth were about 20 dB at 4 kHz and arose up to 35 dB around 10 kHz. However, Mills (1960) had carried out that the minimum threshold for ILDs of pure tones is less than 1 dB. Due to path differences, interaural time differences range from 0 μs for sound incidences from 0° azimuth to about 690 μs for sound incidences from 90° azimuth (Moore, 2003). In contrast to the ILD, these differences vary only slightly with frequency. However, for pure tones the ITD equal interaural phase differences (IPD) to some extent. For low frequencies, these IPD are unambiguous. For high frequencies above 1500 Hz this cue becomes ambiguous as it can no longer be dedicated which cycle in the one ear belongs to a given cycle in the other ear. This ambiguity also holds for narrowband noises. When the IPD is ambiguous, listeners tend to perceive the sound in a location corresponding to the lowest IPD. This effect can be resolved by increasing the bandwidth of the noise as this leads to more or less independent outputs from several auditory filters. Trahiotis and Stern (1989) showed that the required bandwidth for correct localization of a noise with a center frequency of 500 Hz is about 400 Hz.

The assumption that spatial information at high frequencies is caused by ILDs and at low frequencies by ITDs is referred to as "duplex theory" (Rayleigh, 1907). This theory works resonably well for pure tones but is less accurate for complex sounds containing several different frequency components. A click stimulus, e.g., contains energy over a wide range of frequencies. Such a stimulus produces a waveform on the basilar membrane that rather looks like a decaying oscillation. For a click lowpass filtered at 1500 Hz, the relative time or phase difference between the two ears can be compared as the "fine structure" information remains unchanged (Yost *et al.*, 1971). Highpass filtering on the other hand reduces the localization accuracy as only time differences relating the envelope can be compared. In accordance with these findings, Henning (1974) dertermined that the detectability of a 3900 Hz carrier with a 300 Hz modulator is as good as for a 300 Hz pure tone. This holds also for different carrier frequencies at the two ears.

The effectiveness of ITD and ILD alone can be studied using headphones and by presenting the same signal with slighlty different amplitudes or time delays to the two ears. This implies that a shift of the sound to the right side by presenting the sound earlier on that side could be equalized by a higher level on the left side. However, these imprinted differences rather lead to a lateralization than a localization of the signal. Thus, the signal is perceived as being located inside the head and can be projected onto a straight line between the two ears. However,

Plenge (1974) showed that there is only one condition for signals being localized rather than lateralized: the sound signals have to have the same structure as if they were originated from an external sound source, which additionally requires interaural spectral differences.

1.2.2 Head related transfer function

For external sound sources, the upper body, head and pinnae modify the spectra of incoming sounds depending on their angle of incidence. The difference between the spectrum of a sound source and the spectrum of the sound at the position of the ear drum is called head related transfer function (HRTF, e.g. Wightman and Kistler, 1989). It varies systematically with the direction of the sound relative to the head. As stated above, these changes in the spectrum can be used to localize sound sources and thus, lead to a realistic perception of auditory space (e.g. Plenge, 1974). For frequencies above 6 kHz the wavelength is sufficient for interacting with the dimensions of the pinnae. This was confirmed by Gardner and Gardner (1973) who sealed the pinnae cavities and obtained a decrease in localization ability which was largest for broad band and high frequency noises. For lower frequencies the wavelengths interact with the dimensions of head and torso. As the shape of head, torso and pinnae differ across humans, the HRTFs also differ. Wenzel *et al.* (1993) compared the localization ability in free field with the ability of localizing sounds which were presented via headphones using a "representative" HRTF. The results show that cues used for horizontal localization are rather robust against slight changes in the HRTF. However, more errors occured for front-back and up-down discrimination. On the one hand, localization experiments using headphones and head related transfer functions provide reliable and reproducible results independent of the characteristics of the measurement room. On the other hand, the measurement of individual HRTFs is very time consuming and requires special equipment. Thus, Seeber and Fastl (2003) presented a new method for a "subjective selection of non-individual HRTFs". In this method, the most suitable HRTF is singled out of an already existing catalogue of HRTFs in two steps. The selection process comprises six different criteria such as perceiving the sound at a constant distance or elevation. The evaluation by testing the localization ability with the selected HRTF reveals small errors and less fornt-to-back confusion compared to mean or randomly chosen HRTFs.

1.2.3 Static and dynamic angle discrimination

The smallest detectable change in angular position which describes the resolution of the auditory system was, e.g., measured by Mills (1958). He investigated the dependence of the MAA on the reference azimuth and on the frequency of the signal. Results show that with an amount of about 1° the MAA is smallest for a reference azimuth of 0° and frequencies below 1 kHz. In this frequency range, the ITD is the most effective cue. In accordance with the duplex theory, for the same reference azimuth, performance is worst for frequencies around 1500 to 1800 Hz. For higher frequencies, where the ILD is the most effective cue, performance improves again, leading to MAAs around 2°. This MAA rapidly increases with increasing reference azimuth, leading to MAAs not less than 40° over all frequencies for a reference azimuth of 90°.Despite not that pronounced, similar results were shown by Haeusler *et al.* (1983). In normal-hearing listeners, they found MAAs of 1 to 4° for frontal sound incidence and MAAs up to 12° for sounds originating from about ±90°. For hearing impaired listener MAAs ranged from 1 to 10° depending on the type of hearing impairment. For sound incidences from the sides, MAAs were considerably poorer and reached values up to over 30°. They also investigated MAAs in the vertical plane with results from 1 to 6° (3.3° on average) for normal-hearing participants. Similar results for normal-hearing adults (horizontal MAAs of less than 1° and vertical MAAs of less than 4°) were found by Perrott and Saberi (1990). For broadband pink noises Perrott and Pacheco (1989) determined MAAs from 1 to about 4.5°. Additionally, Strybel and Fujimoto (2000) found slightly higher MAAs for short tone bursts of 10 ms compared to bursts with a duration of 50 ms.

The spatial discrimination of sound sources has not only been investigated with static but also with dynamic sound sources. For investigations until the early 1990s, Middlebrooks and Green (1991) have resumed different procedures and results for measurements of the auditory resolution of dynamic sound sources. These studies have typically been performed in two different ways. Listeners have either been asked to discriminate between a static and a moving sound source or to identify the direction of movement. Therefore in most studies physically moving sound sources were used, whereas in some other studies motion was simulated by systematically varying the signal levels between two loudspeakers. Throughout these studies, thresholds have been determined either for stimulus duration, velocity or the change in location. However, all these thresholds have been described as minimum audible movement angles (MAMA). As already shown for the MAAs, MAMAs are also smallest for frontal sound incidence at 0° and increase with increasing azimuth. Grantham (1986), e.g., showed that MAMAs were about 5° for frontal sound incidence and increased to values greater than 30° for stimuli presented from ±90° azimuth. Perrott and Tucker (1988) showed that the performance in detecting MAMAs is worst at frequencies between 1300 and 2000 Hz, i.e. in the same frequency range as for the MAAs, for all tested velocities. As a third parallel between MAA and MAMA, performance decreases with decreasing bandwidth, thus MAMAs are smaller for broadband than for tonal stimuli (Saberi and Perrott, 1990). Additionally, Perrott and Musicant (1977) and

Saberi and Perrott (1990) showed that MAMAs increase with increasing velocity. However, the results of Perrott and Musicant (1977) showed a linear relation between MAMAs and velocity, whereas in the latter study MAMAs remained constant at about 2° for velocities up to 10°/s and then started increasing with increasing velocity. Despite the finding of Perrott and Musicant (1977) which showed that MAMAs cannot be generalized from the localization precision in MAAs, all studies presented above did not find significant differences in threshold for changes in angular displacement for moving and stationary sound sources. What is more: Grantham (1986) found that listeners assess velocity from overall travel distance rather than evaluating the velocity directly, a notion that is known as "snapshot hypothesis". Lufti and Wang (1999) tested this hypothesis by focusing on the detection of angular displacement rather than on how well motion is detected. When expressing their measured thresholds in displacement, acceleration and velocity as total angular displacement, the results were quite similar over all conditions and thus would support the hypothesis. In their task, listeners had to discriminate a change in displacement, velocity or acceleration of a test signal compared to a reference signal. The changes were produced using either one of three different cues: change in intensity, frequency or ITD. An analysis of the cue preference for discrimination, however, did not support the hypothesis. The authors rather determined that for different thresholds different cues were used to discriminate the displacement. Using HRTFs to simulate motion, Carlile and Best (2002) could show that an important factor for the perception of angular movement is the displacement at the endpoint of the motion trajectory.

The acquirement of the measurement procedures and the appropriate results presented so far were used to motivate and discuss the measurement setup and results presented in section 7.

1.2.4 Binaural masking level difference

So far, spatial hearing has been described in terms of differences in sound properties between the two ears for single sound events. Another aspect of spatial hearing is the improvement in detecting signals in presence of a masking stimulus due to binaural differences. This improvement in the signal detection threshold is known as binaural masking level difference (BMLD). When the phase and the level differences at both ears are not the same as those of a masking stimulus, the detectability of the signal improves compared to the case that the differences are equal for the signal and the masker. In practice, this implies that "a signal is easier to detect when it is in a different location than the masker" (Moore, 2003). If the same noise and the same sinusoid were presented at both ears, this leads to a detection threshold L_0 at which the sinusoid is just masked by the noise. The sinusoid becomes audible again if a phase shift of π is applied to the signal. Changing the level of this phase-shifted sinusoid until it is just masked by the noise again leads to the threshold L_π, and thus, L_0 - L_π = BMLD. This difference could gain values up to about 15 dB for signal frequencies around 500 Hz, but decreases to 2-3 dB for frequencies above 1500 Hz. A difference in masking level can also be achieved by presenting a signal and a noise to one ear, adjusting the signal to its threshold and adding the noise to the other ear. With the latter step, the signal becomes audible again, but disappears when the same signal is added to the second side. The different conditions for presentation of masker and noise are indicated using different indices: 0 for the same signal at both ears, π for signal inversion or a phaseshift of π at one ear or m for monaural presentation, and, additionally, u for uncorrelated noises at left and right ear.

The reference condition leading to the threshold L_0 is stated as $N_0 S_0$. Signal-masker configurations that lead to the same

Table 1.1: BMLDs for different conditions for presentation of signal and masker in relation to N_0S_0. Conditions are indicated using the following indices: 0 for the same signal at both ears, π for signal inversion or a phaseshift of π at one ear or m for monaural presentation, and u for uncorrelated noises at left and right ear.

Signal	BMLD [dB]
N_uS_u	3
N_uS_0	4
$N_\pi S_m$	6
N_0S_m	9
$N_\pi S_0$	13
N_0S_π	15

threshold as the reference condition are N_mS_m, N_uS_m and $N_\pi S_\pi$. The effect of BMLD is not restricted to pure tones, the detection threshold of each arbitrary signal can be improved when the phase and level differences of the signal between the ears do not equal those of the masker. As already stated, BMLDs decrease with increasing signal frequency. Higher effects at those frequencies can be achieved by using a narrowband masking noise. This leads to the hypothesis that the release from masking merely depends on the characteristics of the masking noise in the frequency band around the target signal.

1.3 The inner ear and Cochlear Implants

1.3.1 Sound processing in the inner ear

The cochlea is shaped like a spiral with about 2.5 turns and a length of about 32 mm. It consists of three fluid compartments (scala vestibuli, scala media and scala tympani, see upper panel of Figure 1.6) running from the base to the apex. At the base of the cochlea, the scala vestibuli is connected to the ossicles in the middle ear via the round window and the stapes footplate. At the inner tip of the cochlea, the apex, a small opening between the basilar membrane (BM) and the walls of the cochlea, the helicotrema, connects the scala vestibuli and the scala tympani. The BM separates the scala media and the scala tympani. The scala media and the scala vestibuli are connected by a very thin membrane, the Reissner's membrane, and can thus be regarded as one unit (Zwicker and Fastl, 1999). The cochlea is surrounded by bones and filled with two different incompressible fluids. The scala media is filled with endolymphatic fluids that contain positively charged potassium ions. The two remaining scalea contain perolymphatic fluids with low potassium concentration comparable to that in other extracellular fluids (Niparko *et al.*, 2000). Due to the incompressibility, the inward movement of the oval window leads to an outward movement at the round window which closes the scala tympani at the base. This pressure difference forces a movement of the BM leading to a travelling wave from the base to the apex. This motion of the BM according to a sound stimulus is of primary interest for the transformation of a sound stimulus into neural code. The travelling waves start with a small amplitude at the oval window, increase slowly until they reach their maximum at a certain point on the BM and then rapidly die out (Zwicker and Fastl, 1999). Each stimulus frequency causes a maximum displacement at a different region on the BM depending on its mechanical properties that vary from base to apex. The BM is light and stiff at the base and heavier and less stiff at the apex. This change in mass and stiffness is comparable to the strings of musical instruments where heavy strings resonate at high and light strings at low frequencies (Niparko *et al.*, 2000). Similar resonances can be found at the BM: low frequencies cause maximal displacement at the apex, high frequencies at the base. This one-to-one mapping of frequency and location is known as tonotopy (Niparko *et al.*, 2000) or place principle (Zwicker and Fastl, 1999). The frequency that causes the maximum response at a certain location is the characteristic frequency (CF) for that place (Moore, 2003). In addition to the tonotopic organization, the signal processing in the cochlea includes a compressive non-linearity implying that a large range of input levels is compressed to a small range of responses on the BM. The responses are amplified at low and medium levels. This gain decreases with increasing stimulus level, i.e., the response grows more slowly. At very high levels this gain saturates and the response becomes linear. This compression mostly occurs around the maximum of the response and is more pronounced in basal than in apical regions (Moore, 2003).

Inside the scala media, the basilar membrane supports the organ of Corti (see lower panel of Figure 1.6) with its auditory receptor cells (hair cells). The organ of Corti transforms the mechanical oscillations of the BM into electrical signals that can be processed by the nervous system. Due to the ionic environment it already implies electric potential. The stereocilia (rods of protein) at the top of the hair cells have contact with an overlying membrane, referred to as tectorial membrane (TM), inside the endolymph. The bodies of the hair cells, on the contrary, are situated on the BM in the perilymph. The pressure wave along the cochlea moves the basilar membrane relative to the

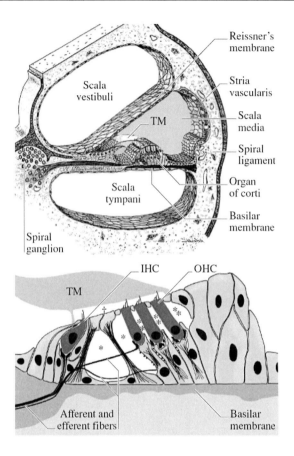

Figure 1.6: **Upper panel**: Cross section of the cochlea showing the scala vestibuli, the scala tympani, and the scala media contaning the tectorial membrane and the organ of corti. **Lower panel**: Scematic representation of the organ of corti, showing the inner and outer hair cells and the tectorial membrane above; from Springer Handbook of Acoustics (2007).

tectorial membrane. This causes a shearing of the stereocilia by splaying them and forcing them together again (Niparko *et al.*, 2000). The stereocilia are connected with the tip links which are stretched during the BM movement and thereby open trans-duction channels for the potassium ions. These depolarize the hair cells by diffusing from the endolymph into the perolymph. This depolarization leads to a release of neuro transmitter that initiate the firing of action potentials in the neuron of the audi-tory nerve. Along the basilar membrane there are differences in structure and function of the hair cells. The bottle shaped inner hair cells (IHC) are arranged in a single row on the inner side of the cochlea. The deflection of their stereocilia causes the release of the neuro transmitter. They are innervated by the auditory nerve fibers and in contact with over 90% of the afferent fibers. Thus they play the principal role in the information conduction to the brain (Niparko *et al.*, 2000). The rod-shaped outer hair cells (OHC) are arranged in three rows on the side closest to the outside of the cochlea. Their lateral walls are free stand-ing and they are closely attached to the tectorial membrane, whereas the IHC have no or only weak contact to the TM. Ad-ditionally, the OHC strongly interact with efferent nerve fibers whereas the IHC rarely receive efferent terminals (Zwicker and Fastl, 1999). As already stated above, the signal transduction is a compressive nonlinear system. At low levels only the OHC are directly stimulated by the shearing force of the tectorial membrane. The deflections of the OHC's stereocilia produce electromotile changes (Wilson and Dorman, 2008) in the length of the cells and thus increase the sensitivity and sharpen the fre-quency selectivity by directly applying energy to the travelling wave (Niparko *et al.*, 2000). At higher levels, also the IHC are directly stimulated by the shearing force of the tectorial mem-brane whereas the OHC are driven to saturation. This leads to the assumption of an interaction between the active OHC and the IHC with the OHC acting as an amplifier with saturating characteristics (Zwicker and Fastl, 1999).

The information about the stimulus is transmitted from the inner ear to the brain through approximately 30,000 afferent nerve fibers. All nerve fibers show spontaneous activity in absence of a stimulus which varies across different fibers. Each has a threshold level below which there is no response and a saturation level above which increasing intensity does not change the response any longer. Thereby, high spontaneous rates are correlated with low threshold and a small dynamic range (Moore, 2003).

Sensorineural hearing loss

The cause of a sensorineural hearing loss can be attributed to disorders in two different areas of the auditory pathway: (i) inside the cochlea (cochlear or sensory hearing loss) or (ii) at the auditory nerve (neural hearing loss). The most common form of sensorineural hearing loss is the damage or lost of the hair cells and thus a disturbance or total disruption of the stimulation of the nerve fibers. In case of damaged inner hair cells, higher thresholds in the related frequency regions are the consequence. The damage of the outer hair cells, on the other hand, leads to a decrease in frequency selectivity which, to a certain extend, disturbes the tonotopic organization (Niparko *et al.*, 2000). The destruction of the hair cells can be caused by genetic disorders, noise exposure, ototoxic drugs, infectuous deseases, Meniere's desease or due to aging. Destructions caused by e.g. loud sounds or ototoxic drugs mostly concern outer hair cells and basal regions which creates a gradient of pathology along the cochlea citepNiparko2000. Without regular activity, the auditory nerve fibers may shrivel. However, even in case of complete hearing loss some fibers retain (Dorman and Wilson, 2004). This survival is greatest for ototoxic infections but decreases with advancing age and long duration of hearing loss (Niparko *et al.*, 2000).

1.3.2 Cochlear Implants

As already stated above, the damage or complete destruction of the hair cells detaches the connection between the basilar membrane and the nerve fibers. Cochlear implants (CI) can, to a certain extend, restore this link by bypassing the missing hair cells and directly stimulating the remaining neurons in the auditory nerve.

The CI consists of five main components, three on the outside and two inside the body (Dorman and Wilson, 2004). The external part comprises (1) an external microphone that pics up the sound and directs it to (2) the speech processor which is situated inside a case behind the ear. This processor transforms the microphone signal into a set of stimuli that can be processed by the electrode array inside the ear. Those signals are conveyed to (3) a transcutaneous coil that transmitts the stimuli as well as the power supply for the inner parts through the skin to (4) the implanted coil. The coil works as a receiver and stimulator and passes the stimuli to (5) an array of electrodes inside the scala tympani. According to manufacturer and type, this electrode array consists of 12 to 22 electrodes arranged on a silicone carrier with an effective length of up to 26 mm and a diameter of less than 1 mm. The single electrodes are connected to the power source via thin platinum wires. There are different types of electrode carriers to cover the individual anatomic conditions, i.e. malformations. Bended carriers for example can be placed closely to the auditory nerve, whereas flexible carriers are used for a placement in the upper windings of the cochlea. In each case different electrodes stimulate different subpopulations of neurons to recreate the tonotopic disposal on the basilar membrane. This spacial specificity not only depends on the number of electrodes but also on different other factors such as the number and distribution of surviving neurons or whether the neural processes in the periphery are present or absent as well as the proximity of the electrodes to the neurons and the electrode coupling configuration (Niparko *et al.*, 2000). Thus, different performances between CI users can be caused by the differences in number and location of surviving cells. This is, as the number of independent stimulation sites is limited by the overlap of the electric fields of different electrodes which depends on the anatomy, e.g., the distance to the target neurons (Wilson and Dorman, 2008).

The main task of the CIs is to transform recorded speech signals into electrical impulses that can, due to the processing, be perceived as speech again. This requires an appropriate coding of the physical characteristics of the (speech) signal such as frequency, intensity or temporal aspects. Due to technical and physiological limits it is not feasible to transmitt the entire information. Thus, there are different methods available for the conduction of signal characteristics:

Continuous interleaved sampling (CIS)

For the CIS strategy the incoming sound is bandpass filtered into a bank of frequencies and the output of each passband is directed to one electrode to imitate the tonotopic frequency order. Subsequently, the envelope of each passband is extracted using low pass filters with cut-off frequencies of 200 Hz or higher to ensure the representation of the fundamental frequencies only in the modulation waveform (Wilson and Dorman, 2008). The envelope variations are represented by modulated trains of biphasic electrical pulses at the corresponding electrodes. The envelope signals are compressed using non-linear filters to map the wide dynamic range of the incoming sound to the smaller electrical dynamic range (Wilson and Dorman, 2008) and are used to control the amplitudes or durations of the stimulation pulses (Wilson *et al.*, 1991). Each of the electrodes is stimulated continuously by the biphasic pulses at a stimulation rate of at least

800 pulses per second (pps). These pulses are interleaved in time across the electrodes to minimize interactions between the electrodes. A close variation of the CIS strategy is the HiRes which uses higher stimulation rates and relatively high cut-off frequencies for the envelope detectors.

Advanced combination encoder (ACE)

The ACE strategy is a channel-selective strategy: Prior to the stimulation, the envelopes for the different channels are scanned to identify the signals with the n highest amplitudes for m different channels. A previous version of this strategy was thus called n-of-m strategy. Subsequently, only the stimulus pulses with the highest amplitudes are directed to the corresponding electrodes. The reason for this feature selection is that in favorable hearing situations, the speech signal should be the loudest signal. In both of the strategies n is set to a fixed value whereas in a third variation of this strategy (spectral peak, SPEAK) can vary from time frame to time frame depending on the level and spectrum of the input signal. The reduced number of stimulating channels still keeps the most important aspects of the signal whereas channels without significant information are excluded. This "unmasking" (Wilson and Dorman, 2008) may emphasize a speech signal against a noise background at least in channels with positive speech-to-noise ratios and thus improve the perception of the principal signal. Except from the channel selection n-of-m and ACE are quite similar to the CIS strategy, even in the stimulation rates with 1000 pps per channel for n-of-m and ACE compared to at least 800 pps for CIS. The SPEAK strategy, in contrast, not only differs in the adaptive selection of the number of channels but with 250 pps also in the stimulation rate.

Fine structure processing (FSP)

The fine structure (e.g. the instantaneous phase or frequency) is an important aspect in speech perception and sound lateralization. The strategies described so far do not seem to comprise a differentiation between envelope and fine structure (FS). However, due to the cut-off frequencies of the low-pass filters substantial FS information is included in the signals and may be perceived at least in the low frequency range (Wilson and Dorman, 2008). One approach to increase the transmission of FS is the FSP strategy. In this strategy, a short group of stimulation pulses is presented at each positive zero crossing in the output of the bandpass filter with the lowest center frequency or, alternatively, up to four bandpass filters with the lowest center frequency (FSP4). Since the length of the pulse sequence is related to the filters upper frequency, the instantaneous repetition rate equals the instantaneous fine structure frequency of the signal (Hochmair *et al.*, 2006). This is in contrast to the continuous representation of pulses for the remaining CIS channels. The overall amplitude, however, is determined by the energy in each channel as in the CIS strategy and the pulses in these channels are also interleaved across electrodes. The advantage of the FSP strategies over the CIS strategy is that a single pulse or short groups of stimulation pulses represent temporal events at least in the lower channels.

1.4 Motivation and structure of the thesis

Despite years of research, the auditory processing of complex signals is not yet fully understood. Thus, a clearer insight into auditory system processes as well as into the relations between physical parameters and hearing sensations still is of great interest. Various investigations address the sound perception at or close to thresholds, while others try to find out whether or how the findings at thresholds imply suprathreshold perception. These relations can for instance be investigated by means of psychoacoustic experiments. One of the keystones of the psychoacoustic sensation is the loudness of acoustic signals. Loudness is a perceptive quantity by which the perception of signals can be described at thresholds as well as at suprathreshold levels. In any way, as described in Section 1.1, it not only depend on the intensity but also on the spectral content and the temporal structure of a signal. Chapters 2 and 3 address the influence of differences in the spectral and temporal structure of a sound on its loudness by investigating (i) non-simultaneous loudness summation using pulse trains with different spectral contents (Chap. 2) and (ii) spectro-temporal weighting of loudness. The former study shows that both repetition rate and pulse duration influence loudness of sequences of tone pulses and that spectral loudness summation is also observed when frequency components are presented nonsimultaneously leading to the assumption that specific loudness builds up rapidly in each auditory channel, but subsides only slowly. Chapter 3 also tries to take the investigation of the loudness of complex, dynamic stimuli one step further. Previous studies either considered the temporal weighting of loudness but did not look at spectral weights, or measured spectral weights but did not consider temporal aspects of loudness. In Chapter 3 spectro-temporal weights for global loudness judgments are estimated by introducing independent temporal variations in level for different spectral regions within each stimulus.

As already mentioned, loudness measurements can be used to determine the absolute threshold of different stimuli and the loudness sensation over the whole auditory dynamic range. They are applicable for the adjustment of hearing aids or cochlear implants (CIs) which are customized by means of hearing thresholds and comfortable levels. Additionally, loudness measurements such as categorical loudness scaling can be used to differentiate between normal and pathological hearing by revealing substantial differences in the shape of loudness functions according to different types of hearing loss, e.g. recruitment. Unfortunately, the results of such loudness scaling experiments (given in categorical units, CU) cannot directly be compared to data derived from loudness models as loudness values are given in phon and sone, respectively. Chapter 4 sheds some light on the relation between the different loudness measures by proposing two functions (differing in the number of parameters) that link the loudness in CU to the classical measure sone.

For the investigations named so far, noise stimuli with different center frequencies or pure tone stimuli are used. Eventhough the sounds that surround us in our everyday life do usually not consist of one signal, but are composed of different signals, e.g. speech in background noise. Fundamental principles of the perception of masked signals are often investigated determining the effects of different noise maskers on sinusoidal stimuli. A so-called release from masking can be observed when (i) the masker has coherent envelope fluctuations in different frequency regions (comodulation masking release, CMR) or (ii), as described in Section 1.2.4, interaural differences between signal and masker are introduced. Such effects are often investigated at threshold levels but do not provide information about perception at suprathreshold levels. Chapter 5 investigates how the suprathreshold perception of a sinusoid masked by bandlimited Gaussian noise is affected due to different temporal structures of the masking noise. Investigations are conducted using categorical loudness scaling, a fast procedure which provides direct information about the shape of the loudness function. The results show that the effects determined at threshold level are also present at suprathreshold levels but decrease with increasing level. Besides these temporal effects, also binaural effects of loudness perception at suprathreshold levels can be investigated using the scaling procedure. Chapter 6 addresses binaural masking effects due to interaural differences of the masking noise in normal-hearing listeners and cochlear implant users.

A different suprathreshold perception is the spatial perception of sounds. On the one hand, this comprises the detection of a certain signal in a number of other signals, e.g. one speaker in a group of speakers. On the other hand, the spatial perception of sounds is defined by the detection or discrimination of different signal positions or moving signals. In Chapter 7 the static and dynamic angle discrimination in free-field is investigated in normal-hearing adults and CI users. As previously presented measurement setups reveal some shortcomings concerning setup related background noise or suitability for cochlea implant users, a new signal generation method is introduced in this chapter. Its feasibility is tested by comparing the normal-hearing results to results of previous studies on this topic. The measured thresholds in static and the dynamic angle discrimination lie within the same range as for previous studies and thus confirm the applicability of the setup. Furthermore, the results show that performance is poorer in CI users but the general tendencies are similar for both groups of listeners.

Modified versions of the Chapters 2 and 4 are published in the "Journal of the Acoustical Society of America" as:

Heeren, W., Rennies, J. and Verhey, J.L. **(2011)** "Spectral loudness summation of nonsimultaneous tone pulses" J. Acoust. Soc. Am. **130** (6) 3905–3915.

Heeren, W., Hohmann, V., Appell, J.E. and Verhey, J.L. **(2013)** "Relation between loudness in categorical units and loudness in phons and sones" J. Acoust. Soc. Am. **130** (6) EL314–EL319.

A modified Version of Chapter 3 is published in "PLOS ONE" as:

Oberfeld, D., Heeren, W., Rennies, J. and Verhey, J.L. **(2012)** "Spectro-Temporal Weighting of Loudness" PLoS ONE **7** (11) e50184.

A modified Version of Chapter 5 is currently under review at the "Journal of the Acoustical Society of America" as:

Verhey, J.L., and Heeren, W. "Categorical scaling of partial loudness under a condition of masking release" (Submission date: 2014-08-18)

In Oberfeld *et al.* (2012) (Chapter 3) and in Chapter 5 the main contributions of the author of this thesis were the conduction of the experiments, part of the data analysis and the data description. In Heeren *et al.* (2013) (Chapter 4) the data were conducted by the third author, Jens E. Appell, whereas the new data analysis and the writing was done by the author of this thesis. In the other studies, the design, conduction and analysis of the data and the writing were done by the author of this thesis.

2 Spectral loudness summation of nonsimultaneous tone pulses[1]

2.1 Introduction

Several studies have shown that loudness increases if the bandwidth of the sound is widened beyond a certain critical bandwidth when the overall sound intensity is kept constant (e.g. Fletcher and Munson, 1933; Zwicker *et al.*, 1957). This effect known as spectral loudness summation has been successfully implemented in loudness models for stationary sounds (e.g. Fletcher and Munson, 1937; Zwicker, 1958; Zwicker and Scharf, 1965; Moore and Glasberg, 1996; Moore *et al.*, 1997; ANSI S3.4, 2007; International Organization for Standardization, 2010). To account for spectral loudness summation, these loudness models commonly assume that the incoming sound is analyzed by a

---

[1]A modified Version of this chapter is published as:

Heeren, W., Rennies, J. and Verhey, J.L. **(2011)** "Spectral loudness summation of nonsimultaneous tone pulses" J. Acoust. Soc. Am. **130** (6) 3905–3915.

bank of overlapping auditory filters (critical bands), processed by a compressive nonlinearity resulting in a specific loudness in each band, and a summation of the specific loudness values across frequency bands.

Studies on the loudness of single bursts have shown that the level of a short signal is higher than the level of an equally loud long signal with the same spectrum (e.g. Munson, 1947; Port, 1963a; Poulsen, 1981; Florentine *et al.*, 1996; Buus *et al.*, 1997; Epstein and Florentine, 2005). This effect is known as temporal loudness summation or temporal integration of loudness and has been successfully accounted for by loudness models assuming that the auditory system integrates the intensity (Poulsen, 1981), neural activity (Zwislocki, 1969), or instantaneous loudness (Zwicker, 1977; Chalupper and Fastl, 2002; Glasberg and Moore, 2002) over time. Zwicker (1969) showed that temporal integration of loudness for single tone bursts could be predicted when a low-pass circuit with a time constant of 100 ms was used (see Chalupper and Fastl, 2002; Rennies *et al.*, 2009). Ogura *et al.* (1991) used the same time constant as Zwicker (1969) to describe the attack of loudness perception, and another, much longer time constant of 5 s to describe the decay. Using the same time constant as Zwicker for attack but a much longer time constant for decay (5 s), Ogura *et al.* (1991) could explain an increase in loudness for a series of repeated noise bursts when the repetition rate was increased. Different time constants for attack and release were also proposed by Port (1963a,b) to account for temporal integration of repeated sound bursts, but the time constants were smaller than those in Ogura *et al.* (1991). Poulsen (1981) argued that a better fit to temporal integration data could be achieved by assuming a serial combination of two leaky integrators with different time constants even for the attack. Glasberg and Moore (2002) published a loudness model with a multiple temporal-integration stage. By computing what they called short-term and long-term loudness, they were able to predict the temporal integration of tone bursts and the loudness of amplitude-modulated signals. Thus, the dynamic properties of these models are expressed in a more or less sophisticated temporal integration stage. This is often the last stage of the models, i.e. effects such as spectral masking or spectral loudness summation are accounted for prior to temporal integration.

Several other studies have investigated effects of spectro- temporal processing in loudness perception. One important finding was that spectral loudness summation depends on stimulus duration (Verhey and Kollmeier, 2002; Fruhmann *et al.*, 2003; Anweiler and Verhey, 2006; van Beurden and Dreschler, 2007; Verhey and Uhlemann, 2008). These studies found an increased spectral loudness summation for short noise bursts (10, 25, or 100 ms) compared to 1000-ms noise bursts. Measuring the level difference between equally loud sequences of broadband and narrowband noise bursts, Verhey and Uhlemann (2008) showed that spectral loudness summation also depends on the repetition rate of short bursts. For low repetition rates, spectral loudness summation was the same as for single short bursts, i.e. larger than for long bursts. For fast repetition rates, spectral loudness summation was similar to that of a single long noise burst. For intermediate repetition rates, Verhey and Uhlemann (2008) observed a continuous transition. In a recent study, Rennies *et al.* (2009) showed that the dependence of spectral loudness summation on duration could not be predicted by models assuming spectral loudness summation prior to temporal integration (Chalupper and Fastl, 2002; Glasberg and Moore, 2002). They suggested a mechanism giving special weight to temporal onsets and showed that such a mechanism could quantitatively predict the dependence of spectral loudness summation on duration as well as on repetition rate. Using perceptual weight analysis, such an onset accentuation in loudness was also found experimentally (Pedersen and Ellermeier, 2008; Rennies and Verhey, 2009; Dittrich and Oberfeld, 2009; Oberfeld and Plank, 2011).

All studies on spectral loudness summation mentioned above used stimuli in which different frequency components were presented simultaneously. In contrast, Zwicker (1969) also considered sequences of short tone pulses with different frequencies of the individual pulses. This sequence had a lower level than an equally loud sequence of the same temporal structure but with pulses of the same frequency. Zwicker (1969) argued that this was related to a specific loudness which built up rapidly in each auditory filter, but decayed only slowly. Such a mechanism would allow spectral loudness summation to take place also for nonsimultaneous tone pulses. This interpretation is difficult, however, based on Zwicker's data alone, because he used pulses with the same level instead of the same loudness. It is therefore likely that the different pulses differed in their loudness and that individual, loud pulses contributed to the increased loudness of the entire sequence. However, Scharf (1970) found nonsimultaneous spectral loudness summation for two consecutively presented equally loud tone pulses with different frequencies indicating that this effect is also observed for tone pulses which are matched in loudness (see also Zwislocki, 1974). In addition, citetZwicker1969 did not systematically measure the effect of the temporal structure of the tone-pulse sequences. If the assumption of a slowly decaying specific loudness is correct, then the nonsimultaneous spectral loudness summation should decrease with increasing inter-pulse intervals (IPIs).

The goal of the present study was to further investigate nonsimultaneous spectral loudness summation for sequences of tone pulses. To facilitate the interpretation of the results, pulses were individually adjusted to the same loudness. Four experiments were conducted: Experiment 1 investigated the role of the repetition rate (i.e. the IPI). If nonsimultaneous spectral loudness summation takes place, a sequence of equally loud tone pulses with different frequencies should be perceived louder than a sequence of pulses with the same frequency, even though the individual pulses are adjusted to the same loudness. This effect should decrease with increasing IPI. For very short IPIs (i.e. for high repetition rates), loudness summation might be reduced, as reported by Verhey and Uhlemann (2008) for simultaneous spectral loudness summation. Experiment 2 investigated the influence of the duration of the individual pulses of the sequences. In experiment 3, the overall stimulus duration was increased to allow for longer IPIs. Experiment 4 was conducted to further investigate the findings of the previous experiments, namely that the loudness of tone-pulse sequences was influenced by both the repetition rate and the pulse duration.

2.2 Methods

2.2.1 Subjects

Twelve subjects with normal hearing participated in the experiments. All had hearing thresholds ≤ 15 dB HL at standard audiometric frequencies between 125 and 8000 Hz, except one who had a threshold of 20 dB HL at 4 kHz. Eleven subjects had experience in loudness measurements. The subjects were between 21 and 29 years old and were paid volunteers except two subjects (S2, S4) who were the first and second author of this paper. All subjects participated in the first two experiments, ten of them also participated in the last two.

2.2.2 Stimuli

The stimuli used in the experiments were sequences of tone pulses. Unless stated otherwise, the overall duration of the stimuli was 1000 ms. The duration of the individual pulses and the repetition rate (i.e. the inter-pulse interval) were varied in the different experiments. The pulses always included 2.5 ms $\cos^2$-ramps at onset and offset. In experiments 1 to 3, the ref-

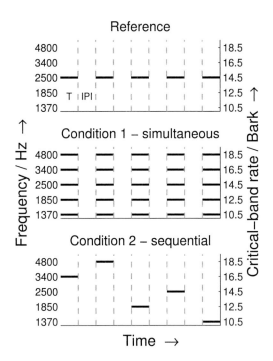

Figure 2.1: Schematic time-frequency representation of the first five
pulses of the stimuli used in the experiments (black horizontal lines). Top: reference stimulus, each pulse has
the same frequency. Middle: stimulus of condition 1,
each pulse contains all five frequencies. Bottom: stimulus of condition 2, each pulse has a different frequency
(in random order). T and IPI denote pulse duration
and inter-pulse interval, respectively.

erence stimulus was a sequence of tone pulses with a frequency of 2500 Hz. Each pulse of the reference stimulus had a level of 60 dB SPL. Two test stimuli were matched in their loudness to the reference signal: (i) a sequence of complex-tone pulses, where each pulse was generated as the sum of the five tone pulses with frequencies of 1370, 1850, 2500, 3400, and 4800 Hz (condition 1), and (ii) a sequence of tone pulses, in which each pulse was a tone pulse of one of these frequencies (condition 2). In condition 2, the sequence of tones was generated by concatenating randomly ordered sequences of the five frequencies until the overall stimulus duration was reached. The five frequencies were separated by 2 Bark on the critical-band-rate scale defined by Zwicker *et al.* (1957) to ensure that spectral loudness summation took place when the frequencies were presented simultaneously. Figure 2.1 shows a schematic time-frequency representation of the first five pulses of the reference stimulus (top) and the test stimuli in condition 1 (middle) and 2 (bottom).

The tones of different frequencies were individually adjusted to the same loudness (see below). The repetition rates of the sequences depended on pulse duration. In experiment 1, 10-ms pulses were used to explore the effect of repetition rate, which ranged between 5 and 100 Hz corresponding to IPIs between 190 and 0 ms (see Table 2.1). In experiment 2, the same measurements were made with pulse durations of 20, 50, and 100 ms. The repetition rates were between 5 and 50 Hz (20 ms), 5 and 20 Hz (50 ms), and between 5 and 10 Hz (100 ms), which corresponded to IPIs between 180 and 0 ms (20 ms), 150 and 0 ms (50 ms), and 100 and 0 ms (100 ms). In experiment 3, nonsimultaneous spectral loudness summation was studied further for 10 ms pulses using a stimulus duration of 2000 ms and IPIs of 0, 90, and 390 ms, i.e. repetition rates of 100, 10, and 2.5 Hz. In experiment 4, the stimuli of condition 1 (simultaneous frequency presentation) and condition 2 (nonsimultaneous frequency presentation) were directly matched in their loudness

Table 2.1: Experimental parameters used in the different experiments of this study: overall duration (Dur), pulse duration (T), inter-pulse intervals (IPIs), and repetition rates (f_{rep}).

Exp.	Dur/ms	T/ms	IPIs/ms	f_{rep}/Hz
I	1000	10	0, 3.3, 10, 30, 90, 190	100, 75, 50, 25, 10, 5
II	1000	20	0, 5, 13.3, 30, 80, 180	50, 40, 30, 20, 10, 5
		50	0, 16.7, 50, 150	20, 15, 10, 5
		100	0, 100	10, 5
III	2000	10	0, 90, 390	100, 10, 2.5
VI	1000	10	0, 3.3, 10, 30, 90, 190	100, 75, 50, 25, 10, 5

by using the test stimuli of condition 1 as the reference. The pulse duration was 10 ms and the repetition rates and IPIs were the same as in experiment 1.

All stimuli were generated digitally using Matlab at a sampling rate of 44.1 kHz. A personal computer controlled stimuli generation and presentation. The stimuli were D/A converted (RME ADI-8 PRO), amplified (Tucker-Davis HB7) and presented diotically to the test subjects via Sennheiser HD650 headphones. The experiment was conducted in a sound-insulated booth.

2.2.3 Procedure

In all experiments the same loudness matching procedure was applied. The loudness of the test signal was matched to that of the reference stimulus using an adaptive two-interval, two-alternative forced-choice procedure. On each trial, the subjects heard two sounds, the reference and the test signal, and indicated which of the two was louder by pressing the corresponding button on a computer keyboard. Test and reference signals were presented in random order and with equal a priori probability. A one-up, one-down procedure tracking the 50%-point of the psychometric function was used (Levitt, 1971). The level of the test signal was decreased if the subject had indicated it to be louder than the reference signal in the previous trial, otherwise it was increased. The initial level increment or decrement was 8 dB; it was halved after each second reversal until a step size of 2 dB was reached. With this step size, the procedure was continued for another four reversals. The mean of the levels of these four reversals was used to estimate the level difference between the reference signal and the equally loud test signal. For each trial, the temporal structure (i.e. pulse duration, IPI, and overall duration) was the same for reference and test signal. The presentation intervals were 1000-ms long in experiments 1, 2, and 4, and 2000-ms long in experiment 3. The two intervals were separated by 500 ms of silence. Three different starting levels of the test signals were used for each track to avoid bias effects: The test signal had an initial level of -30, -20, or -10 dB relative to the reference level in condition 1, and -15, 0, or 10 dB in condition 2. For each condition, the tracks for the different starting levels and IPIs were interleaved to further reduce the influence of bias effects (Verhey, 1999). Thus, the number of tracks depended on the pulse duration with a maximum of 18 (6 IPIs x 3 starting levels) for a pulse duration of 10 ms and a minimum of 6 tracks (2 IPIs x 3 starting levels) for a pulse duration of 100 ms. The measurements were conducted in sessions of two blocks. One block consisted of all tracks (i.e. all IPIs and starting levels) of one condition (simultaneous or non-simultaneous). The other block consisted of all tracks of the other condition (non-simultaneous or simultaneous, respectively). After each session, the subjects could decide whether they wanted to continue with another session after a short break or whether the measurement was continued on another day. For each combination of subject, pulse duration, condition, and IPI the mean level difference across the three starting levels was computed. As in Zwicker (1969), the level differences were always calculated for the frequency component corresponding to the reference frequency (2500 Hz). For example, the complex-tone pulse in condition 1 had an overall level about 7 dB ($= 10 \log(5)$) higher than the reference pure tone pulse at a level difference of 0 dB, provided the five frequency components in the complex tone had the same level.[2]

Prior to the actual experiments, the tones with the different frequencies were equalized in loudness for each subject using the same loudness matching procedure. Although the actual experiments were performed for durations between 10 and 100 ms, the duration of the tone stimuli for the loudness equalization was 500 ms. Poulsen (1981) showed that at a comparable level the temporal integration of loudness was very similar for 1-kHz and 4-kHz tones at durations between 5 and 640 ms. The same time constants were derived under the assumption that two time constants are involved in temporal integration (4 and 100 ms). Thus, for the five frequencies used in the present study, it is reasonable to assume that the level difference at equal loud-

---

[2]In Zwicker (1969), the level of the frequency components of the complex tone were the same and, therefore, the level difference at equal loudness is the same for all frequency components. In contrast, the tones of the present study were adjusted in loudness prior to the experiment resulting in different levels for each component. Thus, the mean level difference applies only to the frequency component of the complex tone at the reference frequency. Since tones with the same loudness were used, the addition of the tones results in a level increase which is slightly different from the 7 dB for equal-level tones.

ness for the sinusoids is independent of duration. Each of the five tones was matched to three different reference tones. All frequencies but the lowest and highest were used as reference frequency. For each reference frequency, three adaptive matches of the test stimulus were conducted with starting levels of -10, 0 and +10 dB relative to the reference level (60 dB SPL). The mean relative level difference between the five frequencies across reference frequencies and starting levels was then used to adjust the tone pulses in the experiments to equal loudness. On average, the standard deviation across subjects, starting levels, and reference frequencies was 1.7 dB indicating that subjects were able to consistently match the loudness of the different frequencies.

2.2.4 Statistical analysis

For each combination of subject, condition, and IPI, the level differences obtained with the three starting levels were averaged. These averages were then used to calculate group averages and standard deviations across listeners. The statistical significance of the measured effects was analyzed by means of analyses of variance (ANOVAs) for repeated measures. The dependent variable was the level difference between test and reference stimulus at equal loudness. The significance level was set to 0.05. Huynh-Feldt corrections were used for the degrees of freedom and the uncorrected degrees of freedom as well as the correction factor ϵ are reported in the results section. Separate analyses were run for each combination of pulse duration, reference spectrum, and overall duration with the fixed factors IPI (two to six levels depending on pulse duration) and condition (two levels: condition 1 and 2). When appropriate, separate ANOVAs were calculated as post-hoc tests for each condition using IPI as fixed factor.

2.3 Results

2.3.1 Experiment 1: Effects of inter-pulse interval

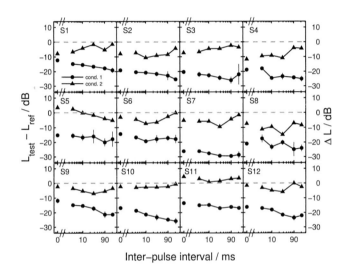

Figure 2.2: Individual data of experiment 1. The level difference between test and reference stimulus at equal loudness is shown for condition 1 (circles) and 2 (triangles) as a function of IPI. Error bars represent standard deviations across the three starting levels.

The individual level differences between reference and test stimuli required for the loudness match are shown in Fig. 2.2 for condition 1 (simultaneous frequency presentation, circles) and 2 (nonsimultaneous frequency presentation, triangles). Error bars denote standard deviations across the three repetitions (i.e. the three starting levels). In condition 1, most subjects showed a similar pattern of level differences, although the individual magnitudes differed. In general, the magnitude of the level difference was largest for long IPIs and decreased with decreasing IPI. All level differences were in the range between -11 and -29.5 dB indicating that the sequence of pulses consisting of five frequencies always required a lower level to be perceived as loud as the reference sequence. Level differences in condition 2 differed considerably from those in condition 1, being always smaller in magnitude. Subjects were more inconsistent in condition 2 than in condition 1: eight subjects showed a general trend characterized by a decrease between IPIs of 0 and 30 ms, followed by an increase for longer IPIs. The other subjects showed increasing (S1, S3), flat (S10), or decreasing (S5) patterns. For all but two subjects (S5, S11), the level differences were always negative indicating that (as in condition 1) the test sequence required a lower level at equal loudness than the reference sequence. The common trends in the individual data were also observed in the mean data as shown in the left panel of Fig. 2.3 (filled symbols). In condition 1 (circles), the magnitude of the level difference increased as the IPI increased, up to an IPI of 90 ms. The same level difference was found for IPIs of 90 and 100 ms. The level differences were between -17 dB at an IPI of 0 ms and -22 dB at 190 ms. In condition 2 (triangles), the level differences were between -2.5 and -6.5 dB and were flatly v-shaped, i.e. a minimum was reached at 30 ms. For smaller and longer IPIs, the magnitudes decreased, reaching about -4 dB at 0 ms and about -2.5 dB at 190 ms. The observations were supported by the ANOVA (Table 2.2). The factors condition and IPI were highly significant ($p < 0.0001$). The significant interaction between condition and IPI indicated that the influence of IPI on the measured level differences depended on condition. Running separate ANOVA for the level differences of each condition revealed that the magnitude of the level differences at the two shortest IPIs (0 and 3.3 ms) were significantly smaller than level differences at the three longest IPIs (30, 90, and 190 ms). For condition 2, the factor IPI was also significant.

2.3.2 Experiment 2: Effects of pulse duration

The second to fourth panel in Fig. 2.3 show mean level differences and standard errors for both conditions for pulse durations of 20, 50, and 100 ms, respectively. The symbols are the same as used for the 10-ms data of experiment 1. The inter-subject variability was similar for all pulse durations. For condition 1 (simultaneous frequency presentation), the comparison across pulse durations showed that the decrease of the level difference with increasing IPI could be observed for all pulse durations. However, the slope of the decrease became increasingly smaller with increasing pulse duration resulting in an almost flat curve for a pulse duration of 100 ms. This was due to the fact that the absolute level differences decreased for long IPIs, while they remained approximately constant for short IPIs. A similar influence of pulse duration on level differences was observed for condition 2 (nonsimultaneous frequency presentation). The v-shaped pattern found in experiment 1 for 10-ms pulses was still observed for pulse durations of 20 ms, although the pattern was not as pronounced as for the 10 ms pulses. For pulse durations of 50 and 100 ms, the curves became increasingly flat. In addition, the absolute level difference at the longest IPI decreased with pulse duration: while a residual level difference of about about 2.5 dB was observed for 10-ms pulses, the level difference was smaller than 1 dB for 100 ms. The ANOVAs (calculated separately for each pulse duration) summarized in Table 2.2 supported these observations. For all pulse durations, the factor condition was significant. The factor IPI was only significant for a pulse duration of 20 ms, reflecting the observed flattening of the curves in Fig. 2.3. A further indication for the increasingly flat curves in both conditions was the fact that the interactions between condition and IPI were significant only for 20 and 50 ms, but not for 100 ms. Separate ANOVAs for each condition and pulse duration showed that, in condition 1, significantly different level differences were found between small and long IPIs. For a pulse duration of 20 ms the magnitude of the level differences at IPIs from 0 to 13.3 ms were significantly smaller than at IPIs of 90 and 180 ms. For a pulse duration of 50 ms the magnitude of the level differences at 0 and 16.7 ms were significantly smaller that at an IPI of 150 ms. No significant level differences were found for a pulse duration of 100 ms. In condition 2, the factor IPI was only significant for a pulse duration of 20 ms, but not for 50 and 100 ms.

2.3.3 Experiment 3: Effect of overall duration

To further investigate the finding that the level difference did not reduce to 0 dB even for the longest IPI using 10-ms pulses, the overall stimulus duration was increased to 2000 ms in experiment 3. Three IPIs were tested with ten of the twelve subjects: 0 and 90 ms as in experiment 1, and 390 ms. The mean level differences are shown in Fig. 2.4 (black). For comparison, the mean data of the ten subjects for experiment 1 are indicated in gray. For both conditions, the level differences at 0 and 90 ms were similar to the data in experiment 1, indicating that overall duration did not affect the level differences at shorter IPIs and that subjects were able to reproduce their results. At an IPI of 390 ms, the level differences were similar to the level differences found at an IPI of 190 ms in experiment 1, i.e. the increase in IPI beyond 190 ms did not affect loudness in either condition. Overall, the data in experiments 1 and 3 were very similar. As in experiment 1, the factors IPI and condition as well as the interaction were found to be significant (top part of Table 2.3).

Table 2.2: Two-way analyses of variance for repeated measures of experiment 1 (10 ms) and experiment 2 (20, 50, and 100 ms). The dependent variable was the level difference between test and reference signal at equal loudness. Condition (Cond; two levels) and IPI (10 and 20 ms: six levels; 50 ms: four levels; 100 ms: two levels) were fixed factors.

	Source	d.f.$_{uc}$	ϵ	Mean sq.	F	Prob > F
10ms	Cond	1	1	8856.9	251.2	≤0.001
	IPI	5	0.58	45.5	8.9	≤0.001
	Cond×IPI	5	0.87	52.2	11.7	≤0.001
20ms	Cond	1	1	7520.7	202.7	≤0.001
	IPI	5	0.58	18.0	4.9	0.007
	Cond×IPI	5	0.68	44.4	12.4	≤0.001
50ms	Cond	1	1	6550.5	341.9	≤0.001
	IPI	3	0.9	0.6	0.3	0.793
	Cond×IPI	3	0.9	15.6	10.0	≤0.001
100ms	Cond	1	1	3453.8	339.0	≤0.001
	IPI	1	1	1.0	0.4	0.521
	Cond×IPI	1	1	3.1	3.0	0.109

Table 2.3: Analyses of variance for repeated measures of experiments 3 (top) and 4 (bottom). The dependent variable was the level difference between test and reference signal at equal loudness. Condition (Cond; two levels) and IPI (three levels) were fixed factors in experiment 3. In experiment 4, IPI (six levels) was the only fixed factor.

	Source	d.f.$_{uc}$	ϵ	Mean sq.	F	Prob $> F$
Exp. 3	Cond	1	1	162.0	313.3	$\leq$0.001
	IPI	2	1	92.1	7.6	0.004
	Cond×IPI	2	0.75	132.3	15.6	0.001
Exp. 4	IPI	5	0.63	74.4	8.9	$\leq$0.001

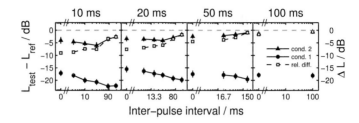

Figure 2.3: Mean data of experiment 1 (left panel) and 2 (other panels). As in Fig. 2.2, the results of condition 1 are indicated with filled circles and those of condition 2 with filled triangles. Error bars represent standard errors across subjects. Open squares represent level differences obtained by subtracting the level differences of condition 1 from the data of condition 2 and shifting the resulting curve vertically to match the level difference of condition 1 for the largest IPI. The different panels show data for different pulse durations indicated at the top of each panel.

2.3.4 Experiment 4: Effect of reference

The data of the previous experiments indicated that the loudness of stimuli with nonsimultaneously presented frequency components (condition 2) was affected by both IPI and repetition rate (see discussion in Section 2.4). To further investigate the interaction of these two factors, experiment 4 was conducted using the same pulse duration, overall duration and IPI as in experiment 1. In the loudness matching procedure, however, the reference was not a single-frequency pulse sequence as in experiments 1 to 3. Instead the sequence of condition 1 with simultaneous frequency components was used as reference signal and the sequence of condition 2 with a different frequency

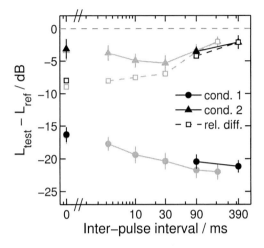

Figure 2.4: Mean data of experiment 3. The data representation is the same as in Fig. 2.3. Gray symbols represent the data of experiment 1.

in each pulse was used as test signal. The experiment thus consisted of a direct loudness equalization of the stimuli used in condition 1 and 2, i.e. the long-term spectrum and the temporal structure of the reference and test stimuli were the same. The level differences at equal loudness (averaged across ten subjects) are shown as black symbols in Fig. 2.5. The positive level differences between about 12.5 and 18.5 dB indicate that stimuli of condition 1 required a lower level to be perceived as equally loud as stimuli of condition 2. With the exception of a single data point the level differences monotonically increased with IPI. This was supported by the ANOVA which showed that the factor IPI significantly influenced the observed level differences (bottom part of Table 2.3).

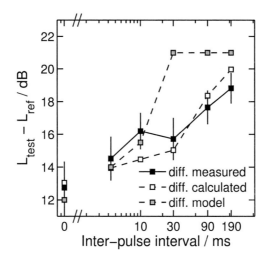

Figure 2.5: Mean data of experiment 4 (black squares). The level difference at equal loudness between sequences of simultaneous (condition 1) and nonsimultaneous (condition 2) frequency components is shown as a function of IPI. Error bars represent standard errors. White squares represent data derived from experiment 1 by subtracting the differences in ΔL relative to the longest IPI in condition 1 from the level differences measured in condition 2 (see left panel of Fig. 2.3). Gray squares represent predictions of the model of Rennies *et al.* (2009).

2.4 Discussion

2.4.1 Spectral loudness summation for pulse trains

In condition 1 of experiment 1, five equally loud tonal components were presented simultaneously. The level difference compared to the equally loud sequences of a single frequency at long IPIs was about 22 dB. About 15 dB of the observed effect was due to spectral loudness summation (see Section 2.2.3). The same pulse duration of 10 ms was also used by Verhey and Uhlemann (2008) to measure spectral loudness summation for sequences of pulses, but with noise instead of tones pulses and IPIs between 0 and 323 ms. Their data, calculated as the level difference between equally loud noise bursts for bandwidths of 200 Hz and 3200 Hz (i.e. the smallest bandwidth they used and the bandwidth corresponding closest to the frequency range covered in condition 1 of this study), are shown together with the present data for sequences of complex tones (condition 1) in the top panel of Fig 2.6. To facilitate the comparison, the level differences at equal loudness of the two data sets are shifted by the magnitude of spectral loudness summation for the longest IPI. Both data sets showed a fast decrease of the level difference as the IPI increased for short IPIs up to about 30 ms followed by an approximately constant level difference for longer IPIs. The data for noise bursts (Verhey and Uhlemann, 2008) were within the standard errors of the level differences for complex tones (present study) for all IPIs that were measured in both studies. Verhey and Uhlemann (2008) argued that the difference between spectral loudness summation for very short IPIs and long IPIs was due to the difference of spectral loudness summation for short signals and long signals. They showed that spectral loudness summation for their longest IPI was the same

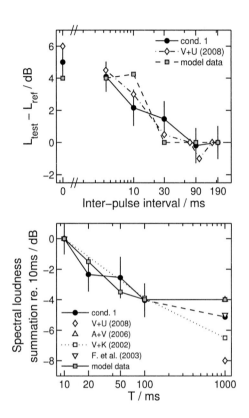

Figure 2.6: Top: level difference between equally loud broadband test signal and narrowband reference signal of condition 1 (shifted upwards by about 22 dB, circles) compared to data measured by Verhey and Uhlemann (2008, diamonds). Gray squares indicate the corresponding predictions of the model of Rennies *et al.* (2009). Bottom: differences in spectral loudness summation as a function of pulse duration T relative to

a pulse duration of 10 ms. Data of condition 1 (circles) are compared to data from the literature (open symbols; diamonds: Verhey and Uhlemann (2008), upward triangles: Anweiler and Verhey (2006), squares: Verhey and Kollmeier (2002), downward triangles: Fruhmann *et al.* (2003)). Gray squares represent the corresponding predictions of the model of Rennies *et al.* (2009).

as for a single 10-ms noise burst and that the magnitude of spectral loudness summation for their shortest IPI was very similar to that for a single 1000-ms noise burst. Their data and the data of the present study indicate that the transition between a spectral loudness summation dominated by a single pulse and that similar to a long signal occurs between an IPI of 0 ms and about 90 ms.

2.4.2 Effect of tone duration in spectral loudness summation

Following the line of arguments of Verhey and Uhlemann (2008), duration effects in spectral loudness summation can also be quantified for the present data set, since spectral loudness summation for the longest IPIs should provide an estimate for the magnitude of spectral loudness summation for a single pulse of the corresponding pulse duration. The bottom panel of Fig. 2.6 shows the change in spectral loudness summation with duration for the present study (filled circles) compared to data in the literature (open symbols). The magnitude of spectral loudness summation was calculated as the difference between the level of the sequence of the complex tones and the reference sequence with pure-tone bursts at the longest IPI. As an estimate for spectral loudness summation of long signals, the level differences at an IPI of 0 ms were averaged across the four pulse durations. This mean level difference is indicated in Fig. 2.6 (bottom, connected by dashed line) at T=1000 ms, i.e. the overall stimulus duration in experiments 1 and 2. The error bar at this data point is the standard deviation across the four pulse durations while, for the other data points, error bars indicate standard errors across subjects. For the data in the literature using noise stimuli, the magnitude of spectral loudess summation was again calculated as the level difference at equal loudness for a bandwidth of 200 Hz and 3200 Hz. To facilitate the comparison between the different studies, spectral loudness summation for a given pulse duration is shown relative to the spectral loudness summation for 10 ms. Almost the same effect of duration was found by Verhey and Kollmeier (2002) and the present study between 10 and 100 ms. There are some differences between the studies in the magnitudes of spectral loudess summation for 10 and 1000 ms, ranging from 4 dB (Anweiler and Verhey, 2006) to 8 dB (Verhey and Uhlemann, 2008). The reason for this difference is unclear but may be due to individual and experimental differences. The estimated magnitude of spectral loudness summation based on the data of the present study for an IPI of 0 ms was about 4 to 5 dB smaller than for the 10 ms pulse with the longest IPI. Thus, the difference in spectral loudness summation for short (10 ms) and long (1000 ms) pulses in condition 1 of this study was similar to that found by Fruhmann *et al.* (2003) and Anweiler and Verhey (2006), but less than found by Verhey and Kollmeier (2002) and Verhey and Uhlemann (2008).

2.4.3 Role of persistence of specific loudness in nonsimultaneous spectral loudness summation

Zwicker (1969) was presumably the first who measured spectral loudness summation for nonsimultaneous presentation of different frequency components. Among different conditions, Zwicker used stimuli with an overall duration of 100 ms composed of five equally long segments. In the sequence with nonsimultaneous frequency components, the segments were 20-ms tone pulses of one of five frequencies (1000, 1350, 1850, 2500, and 3400 Hz), which were separated by 2 Bark on a critical-band-rate scale (i.e. the IPI was 0 ms). Zwicker (1969) found that the level of this stimulus was lower than that of an equally loud sequence in which all segments had a frequency of 1850 Hz.

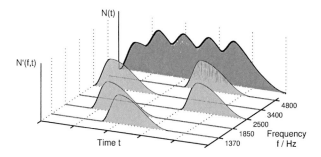

Figure 2.7: Schematic diagram of specific loudness $N'(f,t)$ as a function of time and frequency for a sequence of short tone pulses with different frequencies. In the background, the loudness $N(t)$ is indicated as the sum of $N'(f,t)$ across frequency. The figure was inspired by Fig. 8 in Zwicker (1969).

Zwicker's interpretation of his results is schematically shown in Fig. 2.7, which shows specific loudness as a function of frequency and time in response to a sequence of pulses of different frequencies. Zwicker (1969) argued that specific loudness built up rapidly in each auditory channel, but decayed only slowly (light gray areas in Fig. 2.7). This way a summation of specific loudness across frequency would lead to spectral loudness summation (dark gray area) even if the physical sound presentation was nonsimultaneous. If this hypothesis is correct, the effect of spectral loudness summation for nonsimultaneous frequency components is expected to decrease monotonically with increasing IPI due to the temporal decay of specific loudness. Zwicker (1969) did not, however, systematically vary the IPI to further investigate this effect. One goal of the present study was to test this hypothesis. For 10 and 20-ms pulses in condition 2 of experiments 1 and 2 in this study, a nonmonotonic relation between level differences and IPI was observed. The level differences were largest at intermediate IPIs. This seems to argue against Zwicker's hypothesis. To shed more light on the findings of our experiments, the results of condition 2 were compared to those of condition 1. The underlying reasoning was that data in condition 2 (i.e., for nonsimultaneous spectral loudness summation) could also be influenced by the general dependency of spectral loudness summation on repetition rate (as observed in condition 1 for simultaneous frequency components). Therefore, the level differences measured in condition 1 were subtracted from the level differences in condition 2. Subsequently, the resulting curve was shifted in such a way that the level difference at 190 ms equalled the level difference in condition 2 at the same data point. The results are indicated by open squares in Figs. 2.3 to 2.5 and could be interpreted as the effect of nonsimultaneous spectral loudness summation with removed influence of repetition rate. This effect was measured explicitly in experiment 4, where the two test stimuli of conditions 1 and 2 were directly matched in their loudness. The measured (black squares in Fig. 2.5) and derived level differences (open symbols in Fig. 2.5) are very similar: For all but two IPIs, the derived level differences are within the standard errors of the measurements.

This pattern is similar for all pulse durations (Fig. 2.3). The magnitude of the absolute level difference of the derived curve is largest for an IPI of 0 ms and tends to monotonically decrease with increasing IPI for all pulse durations (with a single exception for the 10-ms pulse). This is in line with Zwicker's hypothesis that a temporal decay of specific loudness is the mechanism underlying nonsimultaneous spectral loudness summation.[3] The combined consideration of conditions 1 and 2 suggests that this effect is partly counterbalanced for short IPIs due to a generally reduced spectral loudness summation for high repetition rates.

2.4.4 Comparison with data in the literature on nonsimultaneous loudness summation

Zwicker (1969) found considerably larger level differences between equally loud single-frequency sequences and sequences of different frequencies (about 12 dB) than in the present study, even though he only tested an IPI of 0 ms, for which a reduced effect was measured in the present study (about 4 dB),

---

[3] Although the data on nonsimultaneous spectral loudness summation are consistent with the assumption of a certain persistence (of specific loudness) after stimulus offset, other mechanisms may also account for the data. For example, some models simulate forward masking on the basis adaptation, reducing the activity after stimulus offset (Dau *et al.*, 1996). Such a mechanism can also predict nonsimultaneous spectral loudness summation if it is assumed that instantaneous specific loudness is derived as a contrast to the activity in other auditory filters. This contrast is enhanced if a preceding stimulus with a different frequency reduces the activity in an off-frequency auditory filter.

presumably due to an influence of reduced overall spectral loudness summation as discussed above. One possible reason is that Zwicker's data were not affected by a considerable reduction in spectral loudness summation for short IPIs since stimuli with an overall duration of only 100 ms were used. Although Verhey and Kollmeier (2002) found a significant difference in spectral loudness summation between 100 and 1000-ms noise bursts for a level comparable to that in the present study, the effect was small (about 2.5 dB) compared to the difference between 10 and 1000-ms bursts (about 6.5 dB). It is reasonable to assume that the difference in spectral loudness summation between stimuli of 20 ms (Zwicker's segment duration) and 100 ms (Zwicker's overall duration) is slightly smaller than between 10 ms and 100 ms (cf. data for 10, 20, and 100 ms in the bottom panel of Fig. 2.6). Therefore, the influence of duration dependence of spectral loudness summation on Zwicker's data can be expected to be smaller than observed in the present data. This may also explain the similar nonsimultaneous spectral loudness summation for the tone-pulse pairs in Scharf (1970) for IPIs of 0 and 200 ms. For an IPI of 0 ms, the total duration of the pair of tone pulses was 10 ms, i.e., the duration of the pulse pair was comparable to the duration of the single pulse (5 ms).

Another possible reason for the larger level difference observed by Zwicker (1969) is that he did not equalize the individual tone pulses with respect to their loudness. Instead, each tonal component had the same level. As shown by Zwicker (1969) the different tones had different loudnesses at the same level. He used 1.85 kHz as a reference in the single-frequency sequence and showed in an additional experiment that, in his study, this frequency was judged least loud among the set of five frequencies. Therefore, the sequence with nonsimultaneous frequency components was louder than the reference sequence at the same level. This certainly contributed to the large level difference in Zwicker (1969).

2.4.5 Residual nonsimultaneous spectral loudness summation

For 10-ms pulses the level difference at the longest IPI was about 2.5 dB. The deviation from 0 dB decreased with increasing pulse duration (see Fig. 2.3). The possible reason that the range of IPIs for 10-ms pulses in experiment 1 was not sufficient to measure the total decay of nonsimultaneous spectral loudness summation was not confirmed in experiment 3. Instead, the level difference remained about the same for IPIs of 190 and 390 ms.

In the light of the hypothesis by Zwicker this would indicate a long-lasting component in the persistence of loudness within each critical band. Zwicker (1984) argued that the decay of forward masking may represent the decay of the internal excitation and thus of the corresponding specific loudness. Although it is commonly assumed that forward masking decays within about 200 ms after masker offset (Moore, 2003), Zeng and Turner (1992) found an increase in just noticeable intensity differences in the presence of the forward masker even 400 ms after forward masker offset indicating an influence of previous stimulation over a range comparable to the present study.

Another possible reason for the residual level difference for long IPIs could be an adaptation to the reference frequency. Ulanovsky *et al.* (2003) showed a decrease of activation of cortical neurons in response to frequently presented spectral components. In condition 2 of the present study, sixty percent of the tone pulses in each trial had the reference frequency. In principle, this could result in a lower loudness of the reference frequency due to stimulus-specific adaptation. The observed duration dependence of the residual level difference at long IPIs, however, seems to argue against this hypothesis.

This duration dependence seems also to argue against explanations on the basis of induced loundess reduction (ILR; for a review see Epstein, 2007). Nieder *et al.* (2003) argued that the psychoacoustical effect ILR may be related to the medial olivo-cochlear reflex (MOCR; for a review see Guinan, 1996). The effect of ILR and MOCR on the stimuli used in the present study is likely to be similar: The response to the pulse in the sequence is reduced due to the presence of the preceding pulses. ILR is frequency selective, i.e., the effect of ILR on the reference signal should be larger than for the test signal where the frequency changes from pulse to pulse. Thus, a lower level of the test signal than of the reference signal is in agreement with the effect of ILR. Since ILR is long lasting it is possible that this level difference is also observed with long IPIs. However, this explanation cannot account for the duration dependence of the residual level difference. The data of Nieder *et al.* (2003) seem to indicate that, for the same duration of inducer and test signal, short signals produce less ILR than long signals whereas the present study found the largest residual level difference for the shortest pulse duration.[4]

2.4.6 Implications for loudness models

The data of the present study show that the temporal structure of sounds affects spectral loudness summation. The findings are in agreement with previous data showing that spectral loudness summation depends on stimulus duration and repetition rate. To our knowledge, the only loudness model which predicts duration dependence of spectral loudness summation was proposed by Rennies *et al.* (2009). Based on the dynamic

---

[4]Since ILR is long lasting, there may also be an ILR between sequences. For example, the test signals in some of the tracks will start at a level of 10 dB above the reference level and ILR is known to be larger for positive level differences of 10 to 20 dB. However, this ILR should equally affect reference and test signals of all other tracks throughout the experimental run. Thus, this ILR would also not account for the residual level difference at long IPIs.

loudness model (DLM) of Chalupper and Fastl (2002), they suggested a bandwidth-dependent amplification of temporal onsets. Rennies *et al.* (2009) showed that such a mechanism could predict duration dependence of spectral loudness summation for single and repeated noise bursts as found by Verhey and Uhlemann (2008). To investigate whether the above conclusions that duration-dependence of spectral loudness summation was one important factor to explain the data of this study, the model of Rennies *et al.* (2009) was applied to the same stimuli. Rennies *et al.* (2009) had fitted the model to the data of Verhey and Uhlemann (2008). In this study, all model parameters were kept the same as originally published. Gray squares in Figs. 2.5 and 2.6 show the resulting predictions. For repeated 10-ms pulses (Fig. 2.6, top), the model predicts the general trend of the data, i.e. a larger level difference is predicted for short than for long IPIs. In contrast to the data, the model does not predict a smooth transition from short to long IPIs, but rather two steady states and an abrupt transition between 10 and 30 ms. For single pulses (Fig. 2.6, bottom), the predictions agree well with the data of different studies showing a smooth decrease of spectral loudness summation with increasing pulse duration up to 100 ms. The fact that no difference is predicted between 100 and 1000 ms is not in line with data of Verhey and Kollmeier (2002), but the difference between 10 and 1000 ms is similar to the data of Fruhmann *et al.* (2003), Anweiler and Verhey (2006), and the present study. Altogether, the model predictions reflect the trends observed in the data and support the above interpretation that duration dependence of spectral loudness summation plays an important role to understand the data of the present study.

In Fig. 2.5, the predictions of the model (gray symbols) are compared to the measured (black symbols) and derived (open symbols) level differences for the stimuli of experiment 4. For IPIs of 0, 3.3, and 10 ms, model predictions agree with the measured level differences. For longer IPIs, predicted level differences are constant at about 21 dB. While this is similar to the measured or derived level difference at the longest IPI, the model does not predict the monotonic increase of level differences for all IPIs. This indicates that the effects of nonsimultaneous across-frequency processing are underestimated for long IPIs. For IPIs larger than 10 ms, the model treats the pulses of different frequencies independently and therefore predicts the same level differences. In the model of Chalupper and Fastl (2002) and consequently also in the model of Rennies *et al.* (2009), two stages contribute to the ability of the model to predict nonsimultaneous spectral loudness summation. The first is a sliding temporal window (Plack and Moore, 1990), which extracts short-term energy of the sound for different frequency bands using a frequency-independent equivalent rectangular duration of 4 ms and a sample rate of 500 Hz. A succession of different frequency components will thus lead to spectral loudness summation if they are separated by less than the duration of the temporal window such as for example the shortest IPI in the present study. The second stage contributing to the effect of nonsimultaneous spectral loudness summation in the DLM and its extended version is a decay of specific loudness in each auditory filter using a nonlinear lowpass, which was motivated by forward masking data and includes a maximum decay time constant of 75 ms. Such a mechanism is a direct realization of Zwicker's suggestion that temporal decay needs to be modeled prior to spectral loudness summation. The comparison in Fig. 2.5 indicates, however, that the modeled decay is too fast to explain the data of the present study. Since Chalupper (2002) adjusted the persistence of specific loudness after stimulus offset based on forward masking data, this may indicate that the effect of persistence of specific loudness is not a direct representation of forward masking (see also Section 2.4.3).

Glasberg and Moore (2002) proposed a loudness model in which temporal decay is modeled after spectral loudness summation, i.e. the decay is not modeled for specific loudness. Their model, however, also allows for nonsimultaneous spectral loudness summation to some extent, since it also applies a short-term analysis of the signals using a time window with a finite length, i.e. nonsimultaneous spectral loudness summation would be predicted by their model if the successive frequency components of the stimulus were not separated by more than the window duration. Unlike the models based on Zwicker's suggestion, however, this effect would depend on frequency since different window durations between 2 and 64 ms are used for different frequency regions (for details, see Glasberg and Moore, 2002). The data of the present study were only collected for a fixed frequency range. Further work is needed to investigate whether the observed effects of nonsimultaneous spectral loudness summation depend on frequency.

To model temporal loudness summation and persistence of loudness, Glasberg and Moore (2002) suggest a two-step integration stage, in which first a so-called short-term loudness is calculated using time constants similar to those in the model of Chalupper and Fastl (2002). Second, a long-term loudness is calculated via temporal integration of the short-term loudness with much longer time constants (attack: 99 ms, decay: 2000 ms). Glasberg and Moore (2002) assumed that the long-term loudness might correspond to a memory for the loudness of an event, which could last for several seconds (see also Oberfeld, 2010). A combination of the decay of specific loudness in Zwicker's approach and the long time constants proposed by Glasberg and Moore (2002) could be used to model the data of the present study. If the interaction of slowly decaying specific loudness and the onset accentuating mechanism suggested by Rennies *et al.* (2009) (or, in fact, any mechanism to predict duration and repetition-rate dependence of spectral loudness summation) is studied further, future models may be able to predict both simultaneous and nonsimultaneous effects of spectral loudness summation.

2.5 Conclusions

1. Both repetition rate and pulse duration influence loudness of sequences of tone pulses. The results of this study are in line with previous results showing that spectral loudness summation is larger for short than for long pulses, and that for low repetition rates loudness summation is similar to that of single pulses.

2. Spectral loudness summation is also observed when frequency components are presented nonsimultaneously. This effect decreases with increasing IPI, but a residual effect at IPIs of 190 and 390 ms remains.

3. Most of the data for nonsimultaneous frequency components can be accounted for by assuming that specific loudness persists after stimulus offset in the corresponding auditory filter while, at the same time, the amount of spectral loudness summation depends on repetition rate. These two effects are additive and partially compensate each other.

4. A loudness model designed to predict duration dependence of spectral loudness summation for single and repeated noise bursts also accounts for the effects of pulse duration and repetition rate for the complex tones of the present study, while data on nonsimultaneous frequency spectral loudness summation require further modeling work.

3 Spectro-Temporal Weighting of Loudness[1]

3.1 Introduction

Loudness is the sensation which is most closely related to the intensity of a sound. However, loudness also depends on other characteristics of the sound such as its duration or its spectral content. Several studies demonstrated that at equal sound pressure level a broadband signal is usually louder than a narrow-band signal (e.g. Fletcher and Munson, 1933; Zwicker *et al.*, 1957; Scharf, 1959; Schneider, 1988; Scharf, 1962; Hübner and Ellermeier, 1993; Cacace and Margolis, 1985; Verhey and Kollmeier, 2002; Verhey and Uhlemann, 2008). This effect is commonly referred to as *spectral loudness summation*. It can be accounted for by assuming that the auditory system analyzes the incoming sound with a bank of overlapping band-pass filters (critical bands) followed by a compressive nonlinearity and a summation across critical bands (Moore *et al.*, 1997; Fletcher and Steinberg, 1924; Zwicker and Scharf, 1965).

Apart from the integration across frequency, the auditory system also seems to integrate over time: The level of a short signal is usually higher than the level of an equally loud long signal

---

[1] A modified Version of this chapter is published as:

Oberfeld, D., Heeren, W., Rennies, J. and Verhey, J.L. (**2012**) "Spectro-Temporal Weighting of Loudness" PLoS ONE **7** (11) e50184.

with the same spectrum (Munson, 1947; Port, 1963a; Poulsen, 1981; Florentine *et al.*, 1996; Epstein and Florentine, 2005). This temporal integration is usually accounted for by assuming a leaky integrator as a temporal integration stage and a decision device that uses the maximum or a percentile of the output of this stage. Current elaborate loudness models include both a spectral and a temporal stage of the kind described above to account for loudness of time-varying sounds like speech or the noise of a vehicle passing by (e.g. Glasberg and Moore, 2002; Chalupper and Fastl, 2002).

The traditional technical measures for the loudness of time-varying sounds, for example L_{Aeq} (the A-weighted energy-equivalent sound pressure level) or N_5 (the 95th percentile of the loudness distribution) (cf. Fastl and Zwicker, 2007; DIN 45631/A1, 2010) which are used in standards on noise assessment (e.g. ISO 1993, 2003), are based on the assumption that all temporal portions of the sound contribute in the same way to overall loudness. More precisely, two temporal portions with identical spectrum and level have the same impact on L_{Aeq} and similar measures, regardless of their temporal position within the sound (e.g., beginning versus end). However, this assumption was recently challenged by studies of temporal weights in loudness judgments. These studies investigated the importance of different temporal segments for global loudness judgments based on stimuli with only small, random level fluctuations (e.g. Ellermeier and Schrödl, 2000; Pedersen and Ellermeier, 2008; Oberfeld, 2008; Dittrich and Oberfeld, 2009). Using methods from so-called *molecular psychophysics* (Green, 1964; Ahumada and Lovell, 1971; Berg, 1989), perceptual temporal weights were obtained at high temporal resolution. The term *molecular psychophysics* (Green, 1964) refers to trial-by-trial analyses that provide information about the relation between a stimulus feature (e.g., the level of one component of a multitone stimulus) and the response of the listener (e.g., "signal present" versus "signal absent"). These methods, for which the alternative terms *perceptual weight analysis* or *behavioral reverse correlation* have been used, typically impose random trial-by-trial variation on the stimulus components. For example, in a study measuring temporal weights (e.g. Oberfeld and Plank, 2011; Rennies and Verhey, 2009), the stimulus might consist of 10 temporal segments. All segments are presented with a mean level of 60 dB SPL, but the levels are randomly and independently varied from trial to trial. Correlational or regression analyses are then used to estimate the impact of the variation of each individual stimulus feature on a behavioral or neural response (for a detailed explanation see (Oberfeld, 2008; Oberfeld and Plank, 2011)). These weights represent the influence of the level of individual temporal portions of a sound on the loudness of the sound as a whole (*global loudness*). More specifically, the weights show how strongly the global loudness changes when the level of a temporal portion of the sound is changed. Studies on the temporal weighting of loudness consistently showed that the first 100-300 ms receive a higher weight than later portions of the stimulus (Pedersen and Ellermeier, 2008; Oberfeld, 2008; Oberfeld and Plank, 2011; Rennies and Verhey, 2009). This means that, for example, a 1 dB increase in the level of the first 100 ms of the sound causes a stronger increase in global loudness than a 1 dB increase in the level of the final 100 ms. This *primacy effect* (highest weight assigned to the beginning of a sound) only occurs for signals with a similar level across the complete duration of the signal (i.e., when the level does not change over time). A *delayed primacy effect* is observed when the level of the first part of the signal is reduced compared to the rest of the stimulus (Oberfeld, 2008; Oberfeld and Plank, 2011). In this case, the first temporal segment presented at the full level receives the highest weight. Results of Oberfeld (2008) suggest that this effect can be explained by attention to the loudest elements, which was proposed as an explanation for higher weights observed for loud elements, even if these elements provided less reliable information than softer elements (Lufti and Jestadt,

2006; Turner and Berg, 2007; Berg, 1990). Some studies also showed a recency effect, i.e., higher weights on the last 100-200 ms of the signal (Ellermeier and Schrödl, 2000; Pedersen and Ellermeier, 2008). However, this effect appears to be considerably smaller than the primacy effect and was non-significant in most studies (e.g. Dittrich and Oberfeld, 2009; Rennies and Verhey, 2009). In any case, the data show that not all temporal portions contribute to loudness to the same extent, contrary to what is assumed by technical measures of loudness such as L_{Aeq} or N_5 (World Health Organization, 1999; European Parliament and Council of the European Union, 2002; Zwicker and Fastl, 1999), which are used in standards on noise assessment like (ISO 1993, 2003). It is interesting to note that, according to simulation results by Pedersen (2006) and unpublished simulations by one author (DO), recent models for the loudness of time-varying sounds (Glasberg and Moore, 2002; Chalupper and Fastl, 2002) (one of them is used in (ISO 1993, 2003)) cannot account for the observed primacy effect, even though they include a temporal integration stage and some effects of temporal masking.

Apart from temporal weights in the processing of the level of the sound, spectral weights have been investigated for sounds without temporal variability. As for the temporal weights, these studies generally reported a non-uniform spectral weighting of auditory intensity (Doherty and Lufti, 1996; Kortekaas *et al.*, 2003; Leibold *et al.*, 2007, 2009; Calandruccio and Doherty, 2008), for example higher weights on the lowest and/or highest frequency component than on the middle components.

Previous studies either considered the *temporal* weighting of loudness but did not look at *spectral* weights, or measured spectral weights but did not consider temporal aspects of loudness. Therefore, in previous experiments, the stimuli were constructed so that there was either only a variation in intensity across time (both within a trial and between trials), or only a variation across frequency (between trials).

Outside the laboratory, however, as real-world sounds like speech or traffic noise unfold in time, the energy in different spectral regions typically evolves differently. In the present study, we therefore estimated *spectro-temporal weights* for global loudness judgments by introducing independent temporal variations in level in different spectral regions within each stimulus. These spectro-temporal weights were compared to temporal weights and spectral weights measured for the same listeners.

3.2 Materials and Methods

3.2.1 Ethics statement

The experiments were conducted according to the principles expressed in the Declaration of Helsinki. All listeners participated voluntarily after providing informed written consent. They received partial course credit or were paid for their participation. The study was approved by the ethical committee of the University of Oldenburg.

3.2.2 Participants

Due to the considerable experimentation time required for each participant (15 sessions, see below), data were collected at two laboratories (Mainz and Oldenburg). In both laboratories, the same software code (MATLAB) was used for stimulus generation and for the control of the experiment. The apparatus was also virtually identical (for details see section Apparatus). In each laboratory, five listeners participated (Mainz: 1 male, 4 female, age 19-29 years. Oldenburg: 2 male, 3 female, 24-31 years). All reported normal hearing and no history of hearing disorders.

3.2.3 Apparatus

The stimuli were generated digitally. In Mainz, they were played back via two channels of a RME ADI/S D/A converter (f_s = 44.1 kHz, 16-bit resolution), attenuated by TDT PA5 attenuators, and buffered by a TDT HB7 headphone buffer. In Oldenburg, they were played back via two channels of a RME ADI-8 PRO D/A converter and amplified by a TDT HB7. In both labs, the sounds were presented diotically via Sennheiser HDA 200 circumaural headphones calibrated according to International Electrotechnical Commission (1970) and free-field equalized according to International Electrotechnical Commission (2009). The experiment was conducted in double-walled sound-insulated chambers. Listeners were tested individually.

3.2.4 Stimuli and conditions

In the *spectro-temporal condition*, the stimuli consisted of 10 temporally overlapping noise segments with a total duration of 120 ms including 20 ms $\cos^2$-ramps at onset and offset (see Figure 1, Panel A). Figure 1, Panel B shows the temporal structure for a single frequency band. The effective duration of each segment was 100 ms. The segments had an overlap of 20 ms resulting in an overall stimulus duration of 1020 ms (effective duration was 1000 ms) . Each temporal segment contained energy in three different frequency bands (see Figure 1, Panel C). The three frequency bands were 3.0 Bark wide to prevent strong level fluctuations that would be audible with smaller bandwidths. In the same vein, we presented Gaussian low-noise noises generated by means of two iterations (Kohlrausch *et al.*, 1997), although this may not have further reduced audible fluctuations because the bandwidth of the stimuli exceeded the auditory filter bandwidth. The noise bands were separated by 4.0 Bark. The noise band with the low center frequency (CF) had cut-off frequencies

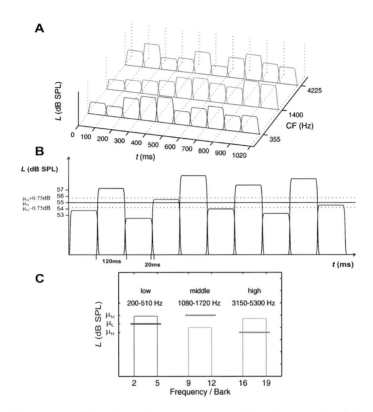

Figure 3.1: Stimulus with spectro-temporal variation. Panel A: In the spectro-temporal condition the stimulus consisted of three narrowband noises. For each noise band, ten temporal segments were presented. Independent and random level perturbations were imposed on the 3 (noise band) × 10 (segment) component levels. Panel B: temporal configuration for the mid-CF noise band. The 10 segment levels were drawn independently from a normal distribution with mean $\mu_M = 55$ dB SPL and

a standard deviation of 2.5 dB. With identical probability, either 0.75 dB was subtracted from or added to
each segment level, in order to create "soft" and "loud"
trials (see text). For the low and high noise band the
same temporal configuration was used. Panel C: Spectral configuration.

of 200 Hz and 510 Hz (2.0 to 5.0 Bark) resulting in an arithmetic
CF of 355 Hz. The cut-off frequencies for the other two noise
bands were 1080 and 1720 Hz (9.0 to 12.0 Bark) for the middle
and 3150 and 5300 Hz (16.0 to 19.0 Bark) for the high frequency
band, corresponding to CFs of 1400 and 4225 Hz, respectively.
The total bandwidth of signals containing all of the three noise
bands was 17.0 Bark.

The middle frequency band was presented at a mean sound
pressure level of μ_M=55 dB SPL. The mean level of the high
(μ_H) and the low frequency band (μ_L) was selected individually on the basis of loudness matches (see Section 3.2.5, sub-
section 3.2.5), so that all noise bands were equally loud. This
is an important aspect of the design because there is evidence
that louder elements receive higher weights (Lufti and Jestadt,
2006; Turner and Berg, 2007; Berg, 1990; Oberfeld, 2009).

On each trial, the sound pressure level of each noise band was
randomly selected for each temporal segment by drawing independently from a normal distribution with mean μ_L, μ_M,
or μ_H and standard deviation 2.5 dB. Thus, in the spectro-
temporal condition the sound contained 30 components (noise
band × segment) with levels independently and randomly selected. On each trial, 0.75 dB was added to or subtracted from
the level of all components (with equal probability) in order to
make it either a "loud" or a "soft" trial, respectively (Oberfeld,

2008). With a 1.5 dB difference between the level distributions of "loud" and "soft" trials we expected the sensitivity in the intensity identification task in terms of the area under the ROC curve (AUC; e.g., Swets, 1986) to be in the vicinity of 0.7, based on previous experience with this kind of task (Oberfeld and Plank, 2011).

The stimuli in the *spectral-weights condition*, the *broadband-noise condition*, and the *single-noise-band conditions* were constructed in the same way as for the spectro-temporal condition, but contained less components. The selection of the mean levels for the three noise bands, the random level perturbations, and the addition or subtraction of 0.75 dB were done exactly as for the spectro-temporal condition. In the *spectral-weights condition* (see Figure 2), there was spectral variation from trial to trial, but no temporal variation within a sound. In this condition, ten overlapping segments each with a duration of 120 ms were presented for each noise band, just as in the spectro-temporal condition. However, within each noise band the levels of the ten segments were identical.

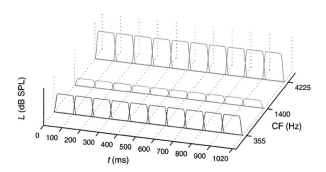

Figure 3.2: Stimulus presented in the spectral-weights condition.

In the *broadband-noise condition* (see Figure 3, Panel A), each segment also comprised the three noise bands, but the levels of the three noise bands were perfectly correlated (r=1.0) in each temporal segment. Thus, effectively a broadband noise varying in level across time was presented, and there was only temporal but no spectral variation (i.e., the spectral composition was identical for all temporal segments, and only the level of the segments varied). Finally, in the three *single-noise-band conditions* (see Figure 3, Panels B to D), each sound consisted of 10 temporal segments containing only the low-CF, middle-CF, or high-CF noise band. Again, there was only temporal but no spectral variation.

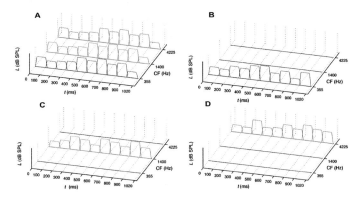

Figure 3.3: Stimuli with purely temporal variation. Panel A: broadband noise. Panel B: low-CF noise band. Panel C: mid-CF noise band. Panel D: high-CF noise band.

3.2.5 Procedures

Loudness matches

Prior to the actual experiment, the three different band-pass noises were equalized in loudness for each listener. For that purpose, the loudness of each noise was matched to the loudness of the other two separately. Within each of the three pairs (low-mid, low-high, and mid-high), both of the noises once served as the reference stimulus (fixed level) and once as the test stimulus (level varied by the adaptive procedure, see below). Additionally, the test stimulus had an initial level of either +10 or -10 dB relative to the level of the reference stimulus, resulting in a total of twelve different adaptive tracks (noise pair × reference stimulus × initial level). To further reduce bias effects, the resulting twelve tracks were randomly interleaved (cf. Buus *et al.*, 1997; Verhey, 1999). Three blocks were presented, each containing the twelve interleaved tracks. For each track, the loudness of the test stimulus was matched to that of the reference stimulus using an adaptive two-interval, two- alternative forced-choice procedure with a one-up, one-down rule (Levitt, 1971), tracking the 50%-point of the psychometric function. On each trial, the listeners heard two sounds and indicated which one was louder by pressing the corresponding button on a computer keyboard. The test stimulus was presented either in interval 1 or in interval 2 with equal a priori probability. The presentation intervals were 1000 ms long and separated by 500 ms of silence. The reference stimulus had a fixed level of 55 dB SPL for the middle-CF noise. For the low-CF and the high-CF reference stimuli, the fixed levels were selected to give the same loudness as the middle-CF reference stimulus according to the loudness model of Chalupper and Fastl (2002). The corresponding levels were calculated as 57.0 dB SPL for the low-CF and 52.0 dB SPL for the high-CF reference stimulus. The level of the test stimulus was decreased if the listener indicated it to be louder than the reference. Otherwise it was increased. The initial increment or decrement in level was 8 dB, and this was halved after each upper reversal until a step size of 2 dB was reached. With this step size, the procedure was continued until another six reversals occurred. The mean of the levels at these final six reversals was used to calculate the level difference between the reference and the equally loud test stimulus. To adjust the noise bands to equal loudness in the main experiment, for each reference noise band the mean level difference between the reference and each test noise band across the two starting levels and the three measurement blocks was computed. Finally, the average of the matches across the three reference noise bands relative to the mid-CF noise band provided the level differences for equal loudness of the three noise bands.

Loudness identification task used for the estimation of weights

A one-interval absolute intensity identification task was used for estimation of the spectro-temporal weights (Braida and Durchlach, 1972). On a four-point ordered rating scale, listeners indicated whether a loud or a soft sound had been presented, and at the same time expressed their confidence when making the decision. The scale comprised the response categories "Soft – rather sure", "Soft – rather unsure", "Loud – rather unsure", and "Loud – rather sure" (in German: "Leise – eher sicher", "Leise – eher unsicher", "Laut – eher unsicher", and "Laut – eher sicher"). Listeners responded by pressing the corresponding buttons on a keyboard. This rating scale including information about confidence was used in order to be able to construct ROC curves for estimating sensitivity. In this way, we avoided the necessity to make potentially unjustified assumptions about the form of the ROC curve, which would be necessary for using d based on binary responses as a measure of sensitivity (Swets, 1986; Macmillan *et al.*, 2004). Because the average overall loudness varied from block to block and was for example higher for the spectro-temporal condition than for the single noise-band conditions, listeners were instructed not to consider the loudness of sounds presented in previous blocks when classifying the sounds in the current block as either soft or loud. Note that the task of deciding whether a soft or a loud trial was presented can be described as a sample discrimination task (Berg and Robinson, 1987; Lufti, 1989; Sorkin *et al.*, 1987). No trial-by-trial feedback was provided.

Each experimental block contained only one of the six conditions (spectro-temporal condition, spectral weights condition, broadband noise condition, and the three single noise band conditions). 91 trials were presented per block. In the main part of the experiment (sessions 5 to 15, see below), each session comprised three blocks of the spectro-temporal condition, and one block of each of the remaining conditions (spectral weights condition, broadband noise condition, and the three single noise band conditions). The order of blocks was randomized, except that the spectro-temporal condition was not presented in consecutive blocks. Across the 11 sessions presenting the intensity identification task, 3003 trials were collected for each listener in the spectro-temporal condition, and 1001 trials in each of the remaining conditions. A higher number of trials was collected for the spectro-temporal condition because of the necessity to estimate a higher number of weights (30) than in the remaining conditions, where only 3-10 weights had to be estimated.

3.2.6 Sessions

In session 1, practice blocks for all experimental tasks and conditions were run. In session 2, loudness matches were obtained for the three noise bands. Additionally, listeners received practice blocks in the identification task. In following sessions only the identification task was presented. Each session lasted about 50 minutes. Listeners participated in one or two sessions per day, separated by a pause of at least 30 minutes.

Weight estimation

The decision weights representing the importance of the 30 spectro-temporal components for the decision in the intensity identification task were estimated from the trial-by-trial data via multiple logistic regression (Pedersen and Ellermeier, 2008;

Oberfeld, 2008; Alexander and Lufti, 2004; Gilkey and Robinson, 1986). For the spectro-temporal condition containing three noise bands with 10 temporal segments each, the decision variable underlying the analysis is given by

$$
\begin{aligned}
D\left(\boldsymbol{L}\right) = {} & \left(\sum_{i=1}^{10} w_{nbL,i} L_{nbL,i}\right) + \left(\sum_{i=1}^{10} w_{nbM,i} L_{nbM,i}\right) \\
& + \left(\sum_{i=1}^{10} w_{nbH,i} L_{nbH,i}\right) - c_j
\end{aligned}
\tag{3.1}
$$

where $\boldsymbol{L}$ is a vector of component levels, $L_{nbL,i}$ denotes the level of the noise band with the low CF in segment i ($i = 1, ..., 10$), $w_{nbL,i}$ is the decision weight assigned to the level of this component, the indices nbM and nbH denote the noise band with the intermediate and the high CF, respectively, and c_j is a constant representing the decision criterion for the jth of the four ordered response categories (Pedersen and Ellermeier, 2008; Berg, 1989; Agresti, 1989). In other words, $D(\boldsymbol{L})$ is a weighted average of the 30 (noise band × segment) independent component levels. The decision model (Eq. 3.1) assumes that, on a given trial, a listener responds that a loud rather than a soft sound was presented if $D(\boldsymbol{L})>0$. More precisely, as we have a four-category response variable Y we assumed a proportional-odds model (McCullagh, 1980) according to which

$$
P\left(Y \leq j\right) = \frac{e^{D(L)}}{1 + e^{D(L)}}, j = 1, ..., J - 1
\tag{3.2}
$$

where J is the number of ordered response categories. This model applies simultaneously to all J - 1 cumulative probabilities, and it assumes an identical effect of the predictors for each cumulative probability (Agresti, 1989). For the remaining conditions, the decision variable was constructed analogously to Eq. 3.1, but contained fewer components. For example, for the spectral weights condition there were only three component levels and three corresponding weights.

In the data analysis, the ordered categorical responses ("Soft - rather sure", "Soft - rather unsure", "Loud - rather unsure", and "Loud - rather sure") served as the dependent variable. The predictors (i.e., component levels) were entered simultaneously. The regression coefficients were taken as the decision weight estimates. For a given component (e.g., the level of the first segment of the low-CF noise), a regression coefficient equal to zero means that the component had no influence at all on the decision to judge the sound as being soft or loud. A regression coefficient greater than zero means that the probability of responding that the *loud* sound was presented increased with the sound pressure level of the given component. A regression coefficient smaller than zero indicates the opposite relation.

Due to the difference in mean level between "loud" and "soft" trials, the component levels were correlated. Therefore, separate logistic regression analyses were conducted for the "loud" trials where 0.75 dB had been added to all component levels, and for the "soft" trials where 0.75 dB had been subtracted (cf. Berg, 1989). This avoids potential problems with multicollinearity, although the multiple logistic regression procedure corrects for correlations between the covariates.

A separate logistic regression model was fitted for each combination of listener, condition, and trial type ("soft" versus "loud" trials). As we were interested in the *relative* contributions of the different components to the decision rather than in the absolute magnitude of the regression coefficients, the decision weights w_i were normalized for each fitted model such that the mean of their absolute values was 1.0 (see Kortekaas *et al.*, 2003), resulting in a set of relative decision weights for each listener, condition, and trial type ("soft" or "loud").

A summary measure of the predictive power of a logistic regression model is the area under the receiver operating characteristic (ROC) curve (Swets, 1986; Agresti, 2002). This measure provides information about the degree to which the predicted probabilities are concordant with the observed outcome (see Dittrich and Oberfeld, 2009, for details). Areas of 0.5 and 1.0 correspond to chance performance and perfect performance of the model, respectively. Across the 120 fitted logistic regression models, AUC ranged between 0.57 and 0.88 (M=0.72, SD=0.08), indicating reasonably good predictive power (Hosmer and Lemeshow, 2000).

3.3 Results

3.3.1 Sensitivity

For each listener and experimental block in the intensity identification task, a ROC curve was constructed from the observed rating response frequencies (for details see Macmillan and Creelman, 2005, Chapter 3). "Loud" trials on which 0.75 dB had been added to all component levels were defined as "signal", and "soft" trials on which 0.75 dB had been subtracted were defined as "noise". The first five trials per block were excluded from the analysis. The AUC was used as an index of sensitivity. AUC does not require strong assumptions about the internal distributions of "signal" and "noise" (Swets, 1986; Macmillan *et al.*, 2004). It corresponds to the proportion of correct responses obtained with the same stimuli in a forced-choice task (Iverson and Bamber, 1997; Green and Moses, 1966) if bias-free responding can be assumed in the forced-choice task (Ulrich and Vorberg, 2009; Yeshurun *et al.*, 2008). To compute AUC, a maximum-likelihood procedure (Dorfman and Alf, 1969) was used for fitting a binormal model (Hanley, 1988). For each block, AUC and its variance were computed from the maximum-likelihood (ML) estimates of slope and intercept of the ROC curve, using the delta method (Metz *et al.*, 1998).

Table 3.1: Mean sensitivity (and its standard deviation) in the intensity identification task, in terms of the area under the ROC curve (AUC), for each condition. Note: *Estimated AUC significantly different from 0.5, p<0.05, two-tailed.

Condition		AUC	SD
Spectral weights		0.68*	0.04
Temporal weights	Low-CF noise band	0.67*	0.07
	Medium-CF noise band	0.64*	0.07
	High-CF noise band	0.64*	0.07
	broadband noise	0.69*	0.06
Spectro-temporal		0.74*	0.07

Table 3.1 shows the average sensitivity (in terms of AUC) for each condition. As expected, the grand mean of AUC was close to 0.70 (M=0.68, SD=0.07). A repeated-measures ANOVA using a univariate approach and Huynh-Feldt correction for the degrees of freedom was conducted (cf. Oberfeld and Franke, 2012), with the within-subjects variable condition (spectro-temporal condition, spectral weights condition, broadband noise condition, and the three single noise band conditions). The df-correction factor $\tilde{\epsilon}$ is reported. Partial η^2 is reported as a measure of effects size. The ANOVA showed a significant effect of condition, $F(5,45)$=19.56, p<0.001, $\tilde{\epsilon}$=0.52, η^2_p=0.69. The highest sensitivity was observed for the spectro-temporal condition. This is compatible with the theory of multiple observations (Green and Swets, 1966), because in the spectro-temporal condition 30 independent components were available, while the remaining conditions contained smaller numbers of components. Post-hoc pairwise comparisons were conducted using non-pooled error terms (Keselman, 1994) and Hochberg's sequentially acceptive step-up Bonferroni procedure (Hochberg, 1988). All except six tests were significant at an α-level of 0.05. The non-significant tests occurred for the three single-noise-band conditions versus the spectral-weights condition, the mid noise band vs. the high noise band, the low noise band vs. the broadband noise, and the spectral-weights condition versus the broadband noise.

3.3.2 Spectro-temporal weights

The normalized decision weights were analyzed via repeated-measures ANOVAs. The results are presented in the following order: 1) the results of the spectral-weights condition where the stimuli varied only spectrally, 2) the results of the conditions showing only temporal variation, and 3) the spectro-temporal weights, and, finally, 4) the weights are compared between some of these conditions.

Spectral weights condition

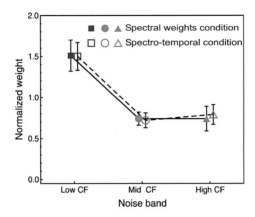

Figure 3.4: Spectral weights. Average normalized relative decision
weights as a function of noise band. Normalization:
mean of the absolute values of the three weights equals
1.0. Filled symbols: spectral weights condition. Open
symbols: spectro-temporal condition, spectral weights
averaged across segments (see text). Error bars show
95% confidence intervals.

A repeated-measures ANOVA with the within-subjects factors
noise band CF (low, middle, high) and trial type (soft or
loud) was conducted to test if the listeners assigned different
weights to the three spectral components (noise bands) in the
spectral-weights condition where there was no temporal variation. As shown by the filled squares in Figure 4, the low-CF
noise band received a higher weight than the two other noise
bands, $F(2,18)=31.41$, $p<0.001$, $\tilde{\epsilon}=0.68$, $\eta_p^2=0.78$. Owing to
the normalization of the weights, the effect of trial type was
not significant. All interactions were also non-significant (all
p-values >0.5).

Temporal weights: single-noise-band and broadband-noise conditions

Rennies and Verhey (2009) found a significantly larger primacy effect for a broadband stimulus than for a narrowband stimulus. To test if such an effect of bandwidth was also observed in the present study an analysis was conducted of the data obtained in the broadband-noise and in the three single-noise-band conditions for which there was only temporal variation. At the same time, the question was addressed if the temporal weights differed between the low, middle and high noise bands. As seen in Figure 5, the weights for all of these four conditions show a clear primacy effect, because the highest weights were observed for segments 1 to 3 (i.e., the first 300 ms of the sound). There was no evidence for a recency effect. The effect of segment was significant, $F(9,81)=68.71$, $p<0.001$, $\tilde{\epsilon}=0.23$, $\eta_p^2=0.88$. The segment $\times$ condition interaction was not significant, however, $F(27,243)=1.14$, $p=0.33$. Thus, the temporal weights did not differ between the three narrowband noises with different center frequencies, and the primacy effect was not significantly stronger for the broadband noise than for the narrowband noises. This effect is in contrast to Rennies and Verhey (2009) who found a significantly larger primacy effect for broadband than for narrowband noise.

There was also a significant effect of condition, and a significant condition $\times$ trial type interaction. Despite the normalization of the weights, these effects appear because some weights with small absolute value were negative. Inspection of the individual data showed that across all listeners and conditions only three of the negative regression coefficients were significantly different from zero. Because in the normalization the mean of the *absolute values* of the weights was set to 1.0, the means were slightly lower for conditions in which negative weights occurred than for conditions where only positive weights occurred. These

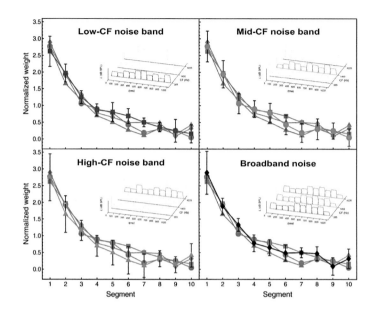

Figure 3.5: Temporal weights. Average normalized relative decision weights for the four conditions showing only temporal variation, as a function of segment number. Normalization: mean of the absolute values of the ten temporal weights equals 1.0. Blue squares: low-CF noise band. Green circles: mid-CF noise band. Red triangles: high-CF noise band. Black diamonds: broadband noise. Error bars show 95% confidence intervals.

effects, however, have no relevance for the interpretation of the data.

Spectro-temporal weights condition

To our knowledge, the spectro-temporal condition including both temporal and spectral variation represents the first report

of spectro-temporal weights in a loudness judgment task. Did
listeners again apply non-uniform temporal or spectral weights,
as in the conditions with only spectral or only temporal variation?

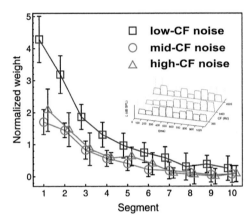

Figure 3.6: Spectro-temporal weights. Average normalized relative
decision weights for the spectro-temporal condition, as
a function of segment number and noise band. Normalization: mean of the absolute values of the 30 (noise
band × segment) weights equals 1.0. Blue squares: low-
CF noise band. Green circles: mid-CF noise band. Red
triangles: high-CF noise band. Error bars show 95%
confidence intervals.

A repeated-measures ANOVA with within-subjects factors segment, noise band, and trial type showed a significant effect of
segment, $F(9,81)=76.74$, $p<0.001$, $\tilde{\epsilon}=0.41$, $\eta_p^2=0.90$. As seen in
Figure 6, the temporal weights for all of the three noise bands
showed a clear primacy effect. The effect of noise band was also
significant, $F(2,18)=35.20$, $p<0.001$, $\tilde{\epsilon}=0.64$, $\eta_p^2=0.80$. As Figure 6 shows, the low-CF noise band received on average higher

weights than the other two noise bands, consistent with the spectral weights observed in the condition showing only spectral variation. The segment × noise band interaction was significant, $F(18,162)=9.36$, $p<0.001$, $\tilde{\epsilon}=0.58$, $\eta_p^2=0.51$. No other effects were significant (all p-values >0.19).

Spectral weights: spectro-temporal condition versus spectral-weights condition

The above analysis showed that the spectral weights for the spectro-temporal condition followed a similar pattern as for the spectral weights condition. The low-CF noise band had a stronger influence on the loudness judgments than the two noise bands with higher CF. As stated in the introduction, an interesting question is whether the listeners applied the same spectral weights in the spectro-temporal condition as in the spectral-weights conditions. In other words, did the presence of temporal variation result in a change in the spectral weights compared to the condition without temporal variation? To answer this question, the means of the absolute values of the weights assigned to the 10 temporal segments for the low, middle, and high noise band in the spectro-temporal condition were computed for each listener and trial type. The resulting three spectral weights were then normalized so that the mean of their absolute values was 1.0. Subsequently, these spectral weights for the spectro-temporal condition were compared to the spectral weights observed for the condition where there was only spectral variation. The average spectral weights estimated for the spectro-temporal condition are shown as open symbols in Figure 4. A repeated-measures ANOVA with within-subjects factors condition, noise band and trial showed a significant effect of noise band, $F(2,18)=37.8$, $p<0.001$, $\tilde{\epsilon}=0.73$, $\eta_p^2=0.81$. The noise band × condition interaction was not significant, $F(2,18)=0.56$, $p=0.56$. Thus, the spectral weights estimated in the spectro-temporal condition did not differ from the weights estimated in the spectral-weights condition. The remaining effects in the ANOVA were not significant (all p-values >0.57).

Temporal weights for the three noise bands: comparison between the spectro-temporal and the single-noise-band conditions

Concerning the pattern of (relative) temporal weights, two questions arise. First, did the temporal weighting patterns differ between the three noise bands in the spectro-temporal condition? Note that the ANOVA presented in the section on temporal weights does not answer this question, because the weights entering this ANOVA represented the effects of both segment and noise band. The second question is: Did the temporal weights for a given noise band estimated in the spectro-temporal conditions show a different pattern than in the corresponding single-noise-band condition?

To answer these two questions, the weights in the spectro-temporal condition were normalized so that the mean of the absolute values of the 10 temporal weights was 1.0 for each listener, noise band, and trial type. The average normalized temporal weights are shown in Figure 7 which compares the weights for the spectro-temporal condition (open symbols) to the weights for the single-noise-band conditions (filled symbols). A repeated-measures ANOVA with within-subjects factors segment, noise band, condition, and trial type showed the expected significant effect of segment, $F(9,81)=84.5$, $p<0.001$, $\tilde{\epsilon}=0.26$, $\eta_p^2=0.90$. The interactions of segment with noise band, condition, or both were not significant (all p-values >0.44). Thus, the pattern of temporal weights did not differ between the single-noise-band conditions and the spectro-temporal condition. There was a significant main effect of noise band, which can again be attributed to spurious negative weights.

A post-hoc ANOVA analyzing only the spectro-temporal condition also showed no significant interaction between segment

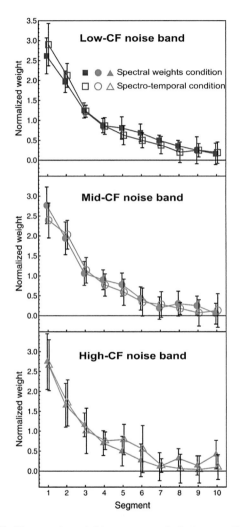

Figure 3.7: Temporal weights compared between the spectro-
temporal condition and the single-noise-band conditions. Average normalized relative decision weights, as
a function of segment number, condition, and noise

band. Normalization: mean of the absolute values of the ten temporal weights equals 1.0 per noise band. Filled symbols: single-noise-band conditions (replotted from Figure 5). Open symbols: spectro-temporal condition (same data as in Figure 6, but different normalization). Error bars show 95% confidence intervals.

and noise band, $\mathbf{F}(18,162)=0.68$, $p=0.81$. Thus, the temporal weights did not differ between frequency bands for this condition, compatible with the comparison between the temporal weights for the single-noise-band and the broadband-noise conditions presented in the section on temporal weights above.

3.4 Discussion

Using stimuli with spectro-temporal variation in level and methods of molecular psychophysics, spectro-temporal weights in a loudness judgment task were obtained. Previous studies estimated either only temporal weights, or only spectral weights. To our knowledge, the only exception is an experiment by Dai and Berg (1992). However, they used a profile listening task where listeners detected a level increment on a single component, rather than performing a global loudness judgment task. The results of the present study demonstrate non-uniform temporal and spectral weights in the spectro-temporal condition.

3.4.1 Comparison to previous studies on temporal weights for broadband stimuli

Several previous studies consistently showed higher weights for the first few segments of a stimulus (primacy effect). In the present experiment higher weights were found for the first three

100-ms segments. This is in the range of two to four segments found in previous studies (Ellermeier and Schrödl, 2000; Pedersen and Ellermeier, 2008; Oberfeld and Plank, 2011). As reported by Rennies and Verhey (2009) and Dittrich and Oberfeld (2009), the weights for the last three segments were very similar to each other. A recency effect, as found by Ellermeier and Schrödl (2000) and Pedersen and Ellermeier (2008), was not observed in the present study. This corroborates the conclusion that the recency effect is considerably weaker than the primacy effect (Oberfeld and Plank, 2011).

It is unlikely that the primacy effect can be attributed to peripheral mechanisms like the initial peak in the firing rate of auditory nerve neurons (cf. Kiang *et al.*, 1965) because it is also observed for a sequence of noise bursts separated by pauses of 5, 40 or 100 ms (for a detailed discussion see Oberfeld and Plank, 2011). With pauses of 40 or 100 ms between the sounds *each* noise burst should have elicited a similar neuronal response, due to the fast recovery of the majority of auditory nerve neurons (Smith, 1977; Harris and Dallos, 1979). Data of Oberfeld and Plank (2011) also argue against a capture of attention due to the abrupt onset of a sound (Pavlov, 1927; Jonides and Yantis, 1988; Sechenov, 1965) as an explanation. They attenuated the abruptness of the onset by imposing a gradual increase in level ("fade in") across the first 300 to 700 ms of a sound with 1 s duration. This did not result in uniform temporal weights, however, but in a delayed primacy effect, with very small weights assigned to the attenuated segments constituting the fade in, and the highest weight assigned to the first unattenuated segment. Dittrich and Oberfeld (2009) proposed that the primacy effect might be caused by a memory process, assuming that the levels of the different temporal portions of a sound are processed as serially sorted information, thus linking the results to experiments on working memory (e.g. Postman and Phillips, 1965) and auditory sensory memory (McFarland and Cacace, 1992), where the characteristic serial position curve also showing a primacy effect is observed (for a detailed discussion see Oberfeld and Plank, 2011). Although it may seem debatable at first sight that the individual segment levels are represented in memory, the assumption that the primacy effect can be attributed to a memory operation rather than to a mechanism specific to auditory intensity processing would be compatible with four observations. First, the primacy effect is observed for a wider range of sensory attributes like intensity (our study), frequency (Berg, 1989), or sound localization cues (Stecker and Hafter, 2002; Stecker and Brown, 2010). This finding could be explained by a higher order mechanism like memory, although it does of course not rule out the possibility that specific mechanisms exist for intensity, frequency, localization etc. which all result in similar temporal weighting patterns. Second, the temporal weights observed for contiguous sounds as used in the present study are very similar to the weights found when the sounds are separated by pauses of 100 ms (Plank, 2005). In the latter case, the stimulus is definitely broken up into perceptually distinct segments. From our own experience as listeners in the task we studied, you can clearly perceive the level changes as a sequence of events, even though the temporal segments are not separated by pauses. Thus, phenomenologically it is conceivable that the temporal portions are processed as separate events. Third, it is important to note that serial position effects like primacy-/recency-effects were not only reported for classical short term memory tasks like remembering a sequence of spoken digits, but also in *sensory memory* (McFarland and Cacace, 1992; Mondor and Morin, 2004). Thus, a "symbolic" representation of the items (in the present example: segment levels) is not a prerequisite for serial position effects. Last but not least, experiments on auditory profile listening where the task was for example to detect a level increment on the temporally central sound in a sound sequence (Oberfeld, 2008; Plank and Ellermeier, 2004) showed that listeners are able to selectively respond to for example a 20 ms segment embedded in a sequence of 10 contiguous segments. This result would be difficult to explain when assuming that listeners have access only to a "unitary" representation of the stimulus, and not to the separate temporal elements.

3.4.2 Bandwidth effects in temporal weights

Rennies and Verhey (2009) argued that the higher weight assigned to the first segment may be due to spectro-temporal effects in loudness. Most studies of temporal weights used broadband stimuli Ellermeier and Schrödl (2000); Pedersen and Ellermeier (2008); Oberfeld (2008); Oberfeld and Plank (2011). Rennies and Verhey (2009) investigated how bandwidth affects the temporal weights. They found that the primacy effect is still present but is significantly reduced when a narrowband signal is used instead of a broadband signal. They argued that this effect of bandwidth may be related to the duration effect in spectral loudness summation. Several studies have indicated that the magnitude of spectral loudness summation is larger for short signals than for long signals. The stronger spectral loudness summation for short signals was attributed to slightly different auditory processing at stimulus onset compared to later portions in time (Verhey and Kollmeier, 2002; Verhey and Uhlemann, 2008; Anweiler and Verhey, 2006; Rennies *et al.*, 2009), resulting in a greater influence of bandwidth on loudness for short than for long signals. According to this hypothesis, the higher weight assigned to the first temporal segment of the broadband signal found by Rennies and Verhey (2009) is due to increased spectral loudness summation at the beginning of the stimulus. Although a duration effect in spectral loudness summation may contribute to the primacy effect, it is unlikely that the primacy effect is solely due to this duration effect, since higher weights are also assigned to segments 2 and 3 (see Figure 5).

The results of the present study do not show a significantly higher weight for the first segment in the broadband condition than for the first segment in the three narrowband conditions. This difference is presumably due to differences in the stimulus parameters. Rennies and Verhey (2009) used a bandpass-filtered noise with a flat spectrum geometrically centered at 2 kHz with a bandwidth of either 400 Hz (from 1810 to 2210 Hz) or 6400 Hz (from 574 to 6974 Hz). The ratio of 16:1 was larger than the bandwidth ratio of the broadband condition and the middle-band-only condition (about 8:1). It is unlikely that the loudness equalization of the bands or the spectral notch between adjacent bands caused the difference in the results between the present study and that of Rennies and Verhey (2009). Heeren *et al.* (2011) found larger spectral loudness summation for a sequence of short complex tones than for long complex tones with the same spectrum, similar to the duration effect of spectral loudness for noise bursts (Verhey and Kollmeier, 2002; Verhey and Uhlemann, 2008; Anweiler and Verhey, 2006). Thus, the effect of duration on spectral loudness summation does not require a continuous spectrum between the lowest and highest frequency components of the stimulus.

3.4.3 Comparison to previous studies on spectral weights

The results of the present study showed a higher weight on the lowest frequency band than on the higher bands (see Figure 4). Such a pattern was not observed in previous studies (Doherty and Lufti, 1996; Kortekaas *et al.*, 2003; Leibold *et al.*, 2007, 2009; Calandruccio and Doherty, 2008), with one exception (Jesteadt *et al.*, 2011). The reason for the different pattern of weights observed in our study may be that the three noise bands were presented at equal loudness. In general, previous studies used the same sound pressure level for all bands. As a consequence, the bands very likely differed in loudness because absolute threshold and loudness are generally frequency dependent (Suzuki and Takeshima, 2004). Below about 4000 Hz, the equal-loudness contours increase towards lower frequencies, i.e., in this frequency range, components with lower frequencies are generally softer than equal-level components at higher frequencies. If one assumes that the loudest components dominate the overall loudness, then the usage of the same SPL may have biased the weights towards the high frequencies. In line with this hypothesis Kortekaas *et al.* (2003) found the highest weight for their highest frequency band, which in their study was often the loudest band (the exceptions are their complex tones with 24 components). The hypothesis is further supported by recent data from Jesteadt *et al.* (2011), who found the highest weight on the lowest frequency component when each component had the same sensation level. In this case, the loudness of the lowest component would have been higher than that of the other components, because the slope of the loudness function is steeper at low than at high frequencies (Suzuki and Takeshima, 2004). It should be noted that Jesteadt *et al.* (2011) also found a slightly higher weight on the lowest band for equal-SPL (rather than equal-SL) bands. This result is difficult to reconcile with the assumption of different weights due to different component loudness. The question of whether the higher weight on the lowest band in the equal-loudness condition can be attributed to the frequency-dependent slope of the loudness function is discussed below (section 3.4.5).

3.4.4 Prediction of spectro-temporal weights from temporal and spectral weights

The analyses presented above show a strong similarity between the average spectral and temporal weights obtained, one the one hand, with the less complex stimuli (i.e., with either only spectral or only temporal variability) and, on the other hand, with the stimuli in the spectro-temporal condition. A simple but strong hypothesis concerning the spectro-temporal weighting of loudness compatible with these results is that the spectral weights do not change as a function of time (i.e., the same spectral weights apply for each temporal segment), and that the temporal weights do not change across frequency (i.e., for each frequency band the same temporal weights apply). If this hypothesis were correct, it would be possible to predict the spectro-temporal weights for each listener by multiplying the spectral weights (estimated in the *spectral-weight condition*) with the temporal weight for each temporal segment (estimated in the *broadband-noise condition*). In more formal terms, the component weights in Eq. 3.1 (e.g., $w_{nbL,i}$) should be given by

$$
\begin{aligned}
w_{nbL,i} &= w_{nbL} \cdot w_i \cdot m, \\
w_{nbM,i} &= w_{nbM} \cdot w_i \cdot m, \\
w_{nbH,i} &= w_{nbH} \cdot w_i \cdot m,
\end{aligned}
\tag{3.3}
$$

where $w_{nbL,i}$, $w_{nbM,i}$, and $w_{nbH,i}$ are the spectral weights for the low-CF, middle-CF, and high-CF noise bands, respectively, as estimated in the spectral weight condition, and w_i is the temporal weight for segment i estimated in the broadband noise condition. The constant m was included because the spectral and temporal weights entering Eq. 3.1 are relative weights and therefore specify the component weights (i.e., regression coefficients) only up to a multiplicative constant.

To test this hypothesis, the individual spectral and temporal weights (as estimated in the spectral weight condition and the broadband noise condition, respectively) were entered into a logistic regression model analyzing the responses from the spectro-temporal condition. As the trial type ("soft" or "loud") had no effect on the weights in the above analyses, the arithmetic mean across the two trial types was used. Thus, for each listener, there were three spectral weights (w_{nbL}, w_{nbM}, and w_{nbH}) and ten temporal weights (w_1 to w_{10}). For a given trial from the spectro-temporal condition, a weighted average Dpred of the 30 component levels presented in this trial was computed, combining Eq. 3.1 and Eq. 3.3. Subsequently, a logistic regression model was fitted relating the response to Dpred. This model had four free parameters: one regression coefficient corresponding to m in Eq. 3.3, and three intercepts corresponding to the cj in Eq. 3.1. For each listener and trial type, the global goodness-of-fit (log likelihood) of this restricted model was compared to the goodness-of-fit of the full model in which the 30 component weights were estimated from the data obtained in the spectro-temporal condition. The full model had 33 free parameters (30 component weights plus 3 intercepts). A likelihood-ratio test was used for model comparison. As the full model has 29 more free parameters than the restricted model, the test statistic is distributed as $\chi^2(29)$. For 10 of the 20 (subject $\times$ trial type) conditions, the fit of the full model was not significantly better than the fit of the restricted model ($p > 0.05$). The average predictive power as indexed by AUC was 0.68 ($SD = 0.05$) for the model based on D_{pred}, which still represents an acceptable quality of the predictions but was significantly smaller than the AUC for the full model ($M = 0.72$, $SD = 0.08$), $F(1,9) = 75.9$, $p < 0.001$. Thus, the 30 component weights for the spectro-temporal condition estimated from the trial-by-trial data for each trial type provided a better fit than the component weights predicted from the spectral and temporal weights (Eq. 3.3). The latter model, however, still provided a surprisingly good fit if one considers that it has 29 fewer free parameters than the full model. It can therefore be concluded that the spectral weights remain constant across time, and the temporal weights do not change across frequency. This pattern of result suggests that the temporal and the spectral weighting of the loudness of a time-varying sound are independent processes. One should note that this may not be true for all stimuli. The bandwidth-dependent weights found by Rennies and Verhey (2009) are not consistent with independent processes. Further studies are needed to investigate for what type of stimuli this assumption is valid.

3.4.5 Spectro-temporal weights within loudness models

As mentioned above, the slope of the loudness function is generally steeper at low than at intermediate and higher frequencies. Suzuki and Takeshima (2004) estimated the slope of the loudness function, i.e. the compressive exponent in the power law relating loudness and intensity to be about 0.3 at 1 kHz and 4 kHz, but about 0.4 at 125 Hz (see their Fig. 7). The relation between the frequency-dependent slope of the loudness function and the spectral weights is consistent with the measured sensitivity for the three single-noise-band conditions. As discussed in section 3.3.1, in those conditions AUC was significantly higher for the noise band with low CF than for the two other noise bands (see Table 3.1).

The optimal decision rule would be to apply spectral weights proportional to sensitivity for the three bands, as indexed by d (integration model; Green, 1958). If one follows this rationale, do the observed differences in sensitivity between the three noise bands account for the observed differences in the spectral weights? To answer this question, d observed in the three single-noise-band conditions (Table 3.1) was individually normalized to an arithmetic mean of 1.0 across the three noise bands. These normalized d values represent the weights predicted by the integration model. Figure 8 plots the observed and predicted spectral weights for each listener. While the observed and predicted weights show rather good agreement for some listeners, there are considerable deviations for other listeners, and in most cases the predicted weights show smaller variability than the observed weights. To quantify the relation between the observed and predicted weights, the correlation coefficient within listeners was computed (Bland and Altman, 1995). This correlation coefficient provides information about whether an increase in one variable (predicted weight) within the individual is associated with an increase in the other variable (observed weight).

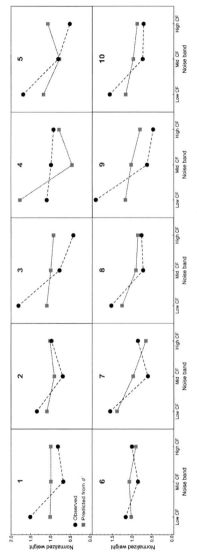

Figure 3.8: Individual spectral weights: observed versus predicted. Average normalized relative decision weights, as a function of noise band. Black circles: weights estimated from the spectro-temporal condition. Gray squares: weights predicted from the sensitivity in the single-noise-band conditions (see text).

The data showed a significant but weak correlation between the observed and predicted weights, $r = 0.44$, $p = 0.039$. Thus, the differences in sensitivity to changes in the level of the three noise bands can partially, but not completely, account for the spectral weights applied by the listeners.

A second interesting question concerning the influence of the slope of the loudness functions on the present data is whether the observed spectral weights can be derived directly from the frequency-dependent exponents. If one assumes that the global loudness is the sum of the component loudnesses (i.e., the component levels in Eq. 3.1 are replaced by component loudnesses), then a level change by for example 1 dB would result in a stronger change in overall loudness when imposed on a stimulus component with a steeper loudness function. As a consequence, the analysis we used would show a higher regression coefficient for a component with a steep slope of the loudness function, that is, a higher relative weight. To answer the question of whether the spectral weights can be explained by the slopes of the loudness function, a power law according to Stevens (1956) and Stevens (1956) was used, representing the most basic loudness model:

$$S = a \cdot p^{2\alpha} \qquad (3.4)$$

where S denotes the perceived loudness, p is sound pressure, a is a dimensional constant, and a is the compressive exponent, i.e. the slope of the loudness function. To focus this analysis purely on the role of the exponent α, any influence of bandwidth was neglected, i.e., the values of a and α as derived for pure tones by Suzuki and Takeshima (2004) were used.

The values of α for the low, middle, and high center frequencies of the stimuli used in the present study were taken directly from Fig. 7 of Suzuki and Takeshima (2004) and were 0.335, 0.293, and 0.289, respectively. The equal-loudness contours shown in Fig. 11 of Suzuki and Takeshima (2004) were used to equalize the loudness of the three frequency components, when the mid-CF component had a level of 55 dB SPL as in the experiment. The corresponding levels of the low- and high-CF components were 57.0 and 51.8 dB SPL. These levels were then used to calculate the dimensional constant a for each CF, such that finally one equation of the form of Eq. 3.4 was available for each frequency component. Subsequently, 5000 random levels were generated for each frequency component using mean values of 57.0, 55.0, and 51.8 dB SPL, a standard deviation of 2.5 dB, and random level increments of +0.75 or -0.75 dB. In other words, the levels were generated exactly as in the experiment except that the frequency components were equalized in loudness based on the equal-loudness contours given in Suzuki and Takeshima (2004). For each component, the loudness was calculated using the power law described above, and the overall loudness was calculated as the sum of the three component loudnesses. This simplifying assumption concerning loudness summation seemed justified because of the relatively wide spacing of the three components (Zwicker and Scharf, 1965). The overall loudness was then used to estimate spectral weights using the same logistic regression procedure as for the data analysis, with a simulated decision criterion corresponding to the arithmetic mean of overall loudness across the 5000 simulated trials. After normalizing to a mean weight of 1.0, this resulted in estimated spectral weights of 1.10, 0.96, and 0.94 for the low-CF, mid-CF, and high-CF components, respectively. In agreement with the data, the weight assigned to the lowest frequency component was higher than for the other frequencies, while the weights on the mid- and high-CF components were similar. However, the difference between the three weights was much smaller than observed in the data (see Figure 4). Thus, the frequency-dependence of slopes of the loudness functions can only partially predict the observed spectral weights. This conclusion is consistent with data of Leibold *et al.* (2007) who measured spectral weights for five-tone complexes with different bandwidths. For bandwidths smaller than 500 Hz, the bowl-shaped spectral weights were rather accurately predicted by the loudness model of Moore *et al.* (1997). However, for a bandwidth of 2119 Hz (which is smaller than the 5100 Hz bandwidth we used), the model predicted almost uniform spectral weights, while the observed weights showed a much higher weight on the highest component than on the other components. Leibold and co-authors suggested that at smaller bandwidths the components mask each other, and that this peripheral interaction between adjacent components is well represented in the loudness model which is based on excitation patterns. In contrast, at the highest bandwidth peripheral interactions should be negligible, so that the non-uniform spectral weights observed for the 2119 Hz bandwidth can be attributed to a more central effect. Similar results were reported by Leibold *et al.* (2009) for a broadband tone complex.

Additional simulations showed that when the three components were identical in *level* (55 dB SPL) instead of being identical in loudness, the estimated normalized weights were 0.87, 0.91, and 1.22. Thus, at equal SPL the highest weight is predicted for the high-CF component. As discussed above, this is consistent with the results of several studies which presented the different frequency components at equal SPL Doherty and Lufti (1996); Kortekaas *et al.* (2003); Leibold *et al.* (2007).

An interesting extension of the present work would be to measure individual loudness functions for single components and spectral weights in the same listeners and to directly analyze the predictability of spectral weights from the measured exponents. Such an analysis could provide a closer view on the contribution of differences in slope to the spectral weights and how individual differences in the slope are reflected in individual differences in the weights. However, the analysis presented above showed that, on average, the higher weight observed for the low-CF component can only partially be attributed to the frequency-dependent slope of the loudness function. Thus, it remains for future research to fully understand the reason for this non-uniform spectral weighting pattern.

At this point, it is important to recall that the weights represent the influence of changes in the level of a specific component on the *global* loudness of the multi-component stimulus. These weights will not necessarily be identical to the slope of the loudness function for a component presented *in isolation*. As an example, consider one of the single-noise-band stimuli (Figures 3 and 5). For example, the slopes of the "isolated" loudness functions are identical for segments 2 and 9, because these components are identical in spectrum, duration, and average level. However, the weights assigned to the two components differ considerably. In other words, the loudness function for a component presented *in isolation* is not identical to the slope of the function relating the component level and the *global* loudness of the multi-component stimulus. The same conclusion can be drawn from the above simulation which demonstrated that the spectral weights cannot be explained by the slope of the loudness function for the three frequency bands presented in isolation.

3.5 Summary and Conclusions

A limitation of previous studies concerned with the decision weights listeners apply when judging the loudness of complex sounds is that they either considered the temporal weighting of loudness but did not look at spectral weights, or measured spectral weights but did not consider temporal aspects of loudness. The present study took the investigation of the loudness of complex, dynamic stimuli one step further and estimated spectro-temporal weights for global loudness judgments by introducing independent temporal variations in level in different spectral regions within each stimulus. The analyses, based on methods of molecular psychophysics, showed that for stimuli which change in spectral composition across time listeners place a higher weight on the lowest frequency component. The temporal weights showed a clear primacy effect, that is, a stronger influence of the sound energy at the beginning of the sound than of the level of later temporal portions. Comparisons of these spectral and temporal weights to weights obtained in control conditions with only temporal or only spectral variation were used to answer the question whether the spectro-temporal weights can be individually predicted from spectral and temporal weights. These analyses showed that this is possible at a rather high precision, indicating that the temporal and the spectral weighting of the loudness of a time-varying sound are independent processes. The spectral weights remain constant across time, and the temporal weights do not change across frequency. The observed non-uniform spectral weights cannot fully be accounted for by loudness models based on the frequency-dependent slope of the loudness function. This observation is compatible with previous suggestions that central rather than peripheral processes are responsible for the observed spectral and temporal weights. These are not yet implemented in current loudness models. Additional research is necessary here.

4 Relation between loudness in categorical units and loudness in phons and sones[1]

4.1 Introduction

Loudness is the sensation of sound that corresponds to the physical sound intensity but that also depends on other sound properties such as frequency content and temporal characteristics. The importance of the sensation of loudness is reflected in the fact that several standards on different aspects of loudness exist (e.g. ISO 226, 2003; ANSI S3.4, 2007; DIN 45631, 1991; ISO 16832, 2006). The present study provides data relating different standardized measures of loudness.

Categorical loudness scaling as standardized in ISO 16832 (2006) provides an easy and fast procedure for measuring loudness over the subject's whole dynamic. Using this procedure the loudness is determined on a scale with named categories such as

---

[1] A modified Version of this chapter is published as:

Heeren, W., Hohmann, V., Appell, J.E. and Verhey, J.L. **(2013)** "Relation between loudness in categorical units and loudness in phons and sones" J. Acoust. Soc. Am. **130** (6) EL314–EL319.

111

inaudible, very soft, soft, medium loud, loud, very loud, and too loud or extremely loud. Loudness in categorical units (CU) has been used in clinical audiology to assess the amount of recruitment in hearing impaired listeners. The procedure has also been used in psychophysics to investigate the temporal integration of loudness (Garnier *et al.*, 1999), spectral loudness summation over a large level range (Anweiler and Verhey, 2006), and as a means for diagnosing suprathreshold hearing loss in individual cases (e.g. Al-Salim *et al.*, 2010).

A limitation of the loudness scaling procedure is that little is known about how loudness in CU relates to the other established measures of loudness, sones and phons. The transformation from CU to sone is, for example, important when data measured with a categorical loudness scaling procedure are compared with predictions of current loudness models, that commonly express their results in sones. The relation between CU and phon may be of interest when relating categorical loudness to established noise emission limits expressed in phons or dBA. To establish this relation, categorical loudness can be compared to standardized equal-loudness-level contours from ISO 226 (2003).

The aim of the present study was to shed some light on the relation between the different loudness measures by comparing categorical loudness data measured according to the ISO 16832 (2006) with the other measures, using the respective standards for phon and sone. Using categorical loudness data for a center frequency of 1 kHz, it was investigated which CU value corresponds to which phon value. It was further investigated if, for normal-hearing listeners, the equal-loudness level contours derived from the categorical loudness data using narrowband noise stimuli are similar to the standardized equal-loudness level contours for tones when the auditory dynamic range is measured with categorical scaling instead of a loudness matching procedure that was used in the studies which formed the basis of ISO 226 (2003). Third-octave filtered noises were used instead of tones, since this is one of the stimulus types suggested in ISO 16832 (2006) for measuring the audible range in Audiology. Furthermore, the influence of the fine structure of hearing threshold (cf. Mauermann *et al.*, 2004) is averaged out when using narrow band noises instead of tones. To relate loudness in CU to loudness in sones, the transformation from phons to sones as tabulated in ANSI S3.4 (2007) was used.

4.2 Methods

The stimuli were third-octave bands of noise with Gaussian amplitude statistics. The center frequencies were 250, 500, 1000, 1500, 2000, 3000, 4000, 6000 or 8000 Hz. The noises were generated digitally at a sampling rate of 44.1 kHz. At the beginning of each run a 5-s long white noise was generated. This noise was transformed to the frequency domain and the magnitudes of the Fourier coefficients were set to zero outside the desired pass band. The signal was then transformed back to the time domain. For each presentation of the stimulus in a run, a 2-s segment was randomly cut out of the 5-s long noise and windowed with 100-ms $\cos^2$ ramps.

Loudness was measured monaurally and separately for each frequency with a non-adaptive loudness scaling procedure according to ISO 16832 (2006). This procedure included two stages. In the first stage, the dynamic range of the listeners was estimated by presenting a sequence of stimuli with increasing level. The listener's task was to press a response button as soon as the stimulus was audible and press it a second time when it was too loud. The maximum level was 120 dB SPL. In the second stage, stimuli were presented twice at each of seven levels that were uniformly distributed over the individual dynamic range as estimated in the first stage. The five named loudness categories in German were "sehr leise" (very soft), "leise" (soft), "mittel" (medium), "laut" (loud) and "sehr laut" (very loud). Four unnamed intermediate response alternatives were represented by horizontal bars. In addition, the two extreme categories "un-hörbar" (inaudible) and "zu laut" (too loud) could be selected, resulting in a total of 11 categories. An eleven-category scale is also used in the example procedure of ISO 16832 (2006). In order to reduce context effects (the tendency of some listeners to rate the current stimulus relative to the previous stimulus), the stimuli were presented in pseudo-random order where the maximum difference of successive presentation levels was smaller than half of the dynamic range of the sequence, in agreement with ISO 16832 (2006).

A personal computer with a coprocessor board (Ariel DSP32C) with 16-bit stereo AD/DA converter was used to control stimulus presentation and recording of the responses of the listeners. The stimulus levels were adjusted by a computer-controlled custom-designed audiometer comprising attenuators, anti-aliasing filters and headphone amplifiers. The signals were presented monaurally to the listeners via headphones (Sennheiser HDA200). The headphones were free field equalized according to ISO 389-8 (2004). The listeners were seated in a sound-insulated booth. Their task was to rate the loudness of each stimulus using a handheld computer (Epson EHT10S) with a LCD touch screen showing the response scale. The handheld computer was connected to the personal computer via a serial interface. The data for both ears of 31 listeners (6 male and 25 female, aged 19 - 34 years) with hearing thresholds below 10 dB HL for frequencies between 125 Hz and 4 kHz and below 15 dB HL for 6 to 8 kHz were analyzed following the recommendations of the ISO 16832 (2006). For this, loudness categories were linearly transformed to numerical values (categorical units, CU) from 0 (inaudible) to 50 (too loud). A model loudness function (i.e., CU vs. level) was fitted to the individual data for each frequency as described by Brand and Hohmann (2002). The function consists of two linear parts with independent slope values and a transition region which was smoothed using a Bezier fit. The function was fitted to the individual data using a modified least-square fit according to Brand and Hohmann (2002). Using the fitted functions, the levels corresponding to given loudness values in CU were derived from the individual loudness functions for each frequency. These individual loudness levels were then sorted in ascending order for each frequency and CU value separately and the median was determined. The median loudness function for the group of subjects was then determined by fitting a loudness function to the median data, using the same function as for the individual data. This procedure is in agreement with the suggestion for the determination of average loudness functions in section 5.1 of the ISO 16832 (2006).

4.3 Results and Discussion

Figure 4.1 shows the median loudness functions (loudness in CU versus level in dB SPL) for each center frequency (black solid line). All functions show a steeper slope for higher than for lower levels. The inter-quartile range is shown as the red area and the blue area indicates the range between the 5% and the 95% percentile. The shapes of the loudness functions and the percentile range are comparable to the loudness functions shown in Appendix A of ISO 16832 (2006).

The following considerations are based on the assumption in ISO 226 (2003) that the loudness of a third octave noise is the same as for a tone at the center frequency of the noise. This assumption is required to allow for the comparison between the loudness scaling data measured with third-octave band noises and the standardized loudness measures in sone and phon, which are based on results of experiments with pure tones.

Equal-loudness-level contours corresponding to 3, 20, 40, 60, 80 and 100 phon were derived from the median loudness-scaling data (black lines in Figure 4.1) in two steps. In the first step, the loudness in CU of the noise centered at 1000 Hz was de-

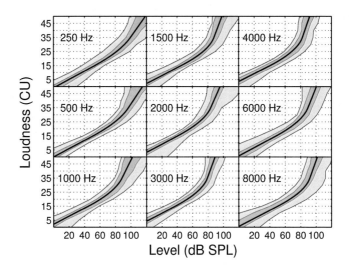

Figure 4.1: Median loudness function across 62 ears (solid black
lines) obtained with the categorical loudness scaling
procedure, interquartile range (red area), and maximum range (blue area), for the nine center frequencies.

termined for levels of 3, 20, 40, 60, 80 and 100 dB SPL using the median loudness function, resulting in CU values that
correspond to 3, 20, 40, 60, 80 and 100 phons. In the second step, the levels in dB SPL were derived from the loudness
functions for the other center frequencies for the respective CU
values. Figure 4.2 shows the resulting equal-loudness-level contours. The solid red lines indicate the equal-loudness-level contours derived from the median categorical loudness functions,
and black dashed lines show those given in ISO 226 (2003).
The curves match at 1000 Hz, since data at this frequency were
used as the reference points. The differences between the equal-
loudness-level contours for center frequencies up to 6000 Hz and

loudness levels up to 60 phon are smaller than 5 dB except for the 3-phon equal-loudness-level contour at 500 Hz. For parts of the 80-phon equal-loudness-level contour, for all equal-loudness-levels at 8000 Hz, and for the equal-loudness level at 3 phon for 500 Hz, the differences are between 5 and 7.5 dB. For frequencies below 1 kHz of the 100-phon contour, differences increase up to 14 dB at 250 Hz. For frequencies above 1 kHz no data are provided in ISO 226 (2003) for 100 phon and thus the measured data cannot be compared to the standard. In summary, differences are small at low and intermediate levels and at medium frequencies and larger at high levels and the highest and lowest frequencies.

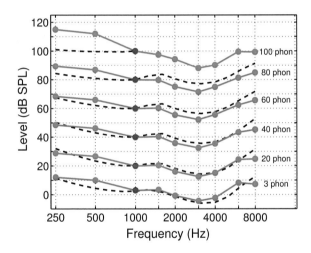

Figure 4.2: Equal-loudness level contours on the basis of loudness scaling data (solid red lines) and according to ISO 226 (2003) (dashed black lines).

One explanation for the differences between the standardized equal-loudness-level contours and those derived from the CU data may be that the accuracy of the categorical scaling procedure is limited. Recently, Al-Salim *et al.* (2010) measured loudness functions according to ISO 16832 (2006) in two separate sessions to investigate the reliability of categorical loudness scaling. They found that the slopes of the loudness functions for lower levels were essentially the same for the two sessions whereas the slopes for higher levels were not reliable across sessions. This may account for the differences between the ISO 226 (2003) and the equal-loudness-level contours derived from the CU data at the highest loudness levels. An alternative explanation is that the shapes of the standardized equal-loudness-level contours are influenced by the procedure used to derive them. Using a constant stimulus procedure, Gabriel *et al.* (1997) showed that the shapes of equal-loudness-level contours depend on various experimental parameters, e.g., the level range and the frequency difference between the reference (usually a 1000 Hz tone) and the test stimulus. Their comparison of the test-tone levels of different studies revealed that, using a constant stimulus procedure, the absolute position of the level range of the test stimuli strongly affected the shape of the equal-loudness-level contours. Increasing the average level of the range increased the estimated equal-loudness level. This range effect increased as the difference between the test and reference frequency (1000 Hz) increased and was, e.g., up to 12 dB for a test-tone frequency of 125 Hz at 30 phon. Such effects should not occur with categorical scaling since signals with different frequencies are not directly compared. Thus one may argue that unbiased equal-loudness level contours may be closer to the ones derived from loudness scaling data than those derived from data where signals with different frequencies are compared. Following this line of argument, the data presented here may provide interesting information about the shapes of the equal-loudness-level contour at very high levels where the data in the literature are sparse and therefore may be more sus-

ceptible to biases due to specific experimental parameters used
in those few studies.

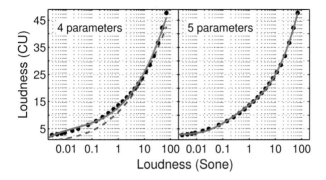

Figure 4.3: Cubic-fit function relating loudness in sones to loudness
in CUs at 1kHz using four parameters (left) and five
parameters (right). The black dots represent median
subject data at 1000 Hz, and the dashed gray line shows
a cubic-fit function proposed by Appell (2002) that was
used in Anweiler and Verhey (2006).

To examine the relation between loudness in CU and loudness
in sones, the median loudness function for a center frequency
of 1000 Hz was compared to the loudness in sones. For the
CU values, the loudness in sones was determined using the relationship between loudness level in phons and loudness in sones
tabulated in ANSI S3.4 (2007). For this purpose, loudness values in CU were determined for loudness levels given in the ANSI
standard (1, 2, 3, 4, 5 and 7.5 phon and between 10 and 100
phon in 5-phon steps) using the median loudness function for
1000 Hz of the present study (black line of the 1000-Hz panel
in Figure 4.1). Figure 4.3 shows the loudness in CU as a function of the loudness in sones for a center frequency of 1000 Hz
(black dots). As proposed in a previous study by Appell (2002),

a cubic relation between loudness in sones and loudness in CU was assumed. Although such an equation cannot be inverted analytically, the CU values may be converted to sones using a numerical approach on the basis of the given equation. Two fit functions are proposed which differ in the number of parameters. The left panel of Figure 4.3 shows a cubic fit function with a solid red line analogous to Appell (2002), Figure 5.5 using four parameters:

$$N_{CU} = 0.7495 \cdot lg\left(N_{sone}\right)^3 + 3.6827 \cdot lg\left(N_{sone}\right)^2 \\ + 8.6284 \cdot lg\left(N_{sone}\right) + 12.8530 \tag{4.1}$$

The parameters differ from the previous fit by Appell (2002) as the underlying datasets are slightly different. For comparison, his fit function is shown with a dashed gray line. For most loudness values, the new fit function predicts the relation between CU and sone with high accuracy (root mean square error 0.7524 CU). An even better fit can be obtained when a fifth parameter is introduced allowing for a shift on the logarithmic sone scale:

$$N_{CU} = 2.6253 \cdot lg\left(N_{sone} + 0.0887\right)^3 \\ + 0.7799 \cdot lg\left(N_{sone} + 0.0887\right)^2 \\ + 8.0856 \cdot lg\left(N_{sone} + 0.0887\right) + 13.4493 \tag{4.2}$$

This shifted cubic fit function is shown in the right panel of Figure 4.3. The root mean square error (0.7524 CU) was reduced to 0.3016 CU. One should note that the transformation is based on a comparison of two data sets, i.e., as for the equal-loudness-level contours one cannot rule out that differences in the group of subjects result in slightly different slopes.

In summary, the present study presented data that linked the loudness in categorical units to the classical measures of loudness sone and phon. These relations may prove to be useful when loudness growths function of other sounds such as environmental sounds are assessed with this fast standardized procedure, e.g., in comparison to predictions of loudness models.

5 Categorical scaling of partial loudness under a condition of masking release[1]

5.1 Introduction

Categorical loudness scaling is a standardized procedure to measure loudness growth functions in audiology (ISO 16832, 2006). The present study investigated to what extent categorical scaling can be used as a fast procedure to assess partial loudness of a masked signal under a condition of masking release.

Masking experiments are often used to investigate characteristics of auditory processing, such as its frequency selectivity (e.g., Patterson, 1976). This data is often interpreted on the basis of the assumption that the auditory system analyzes the energy of the sounds. A release from masking usually provides insights into processes beyond such a simple analysis of the energy of the target signal and the masker. This is observed when,

---

[1] A modified Version of this chapter is currently under review at the "Journal of the Acoustical Society of America" as: Verhey, J.L., and Heeren, W. "Categorical scaling of partial loudness under a condition of masking release" (Submission date: 2014-08-18)

e.g., interaural disparities between signal and masker are introduced resulting in a binaural masking level difference (BMLD) or when the masker has coherent envelope fluctuations in different frequency regions. This latter effect is commonly referred to as comodulation masking release (CMR, Hall *et al.*, 1984; Verhey *et al.*, 2003, for a review). While the BMLD experiments provide information about the processing of the two ear signals, CMR experiments indicate how modulation is processed within and across different auditory channels. A limitation of studies focusing on masked thresholds is, however, that they do not necessarily provide information about perception at suprathreshold levels. This information may be used to test auditory models. In addition it is important when physiological correlates of masking are investigated where the signals are often presented at suprathreshold levels (e.g., Sasaki *et al.*, 2005; Ernst *et al.*, 2010).

Only a few studies investigated how the loudness of a masked signal is altered when cues eliciting a masking release were introduced (Townsend and Goldstein, 1972; Soderquist and Schilling, 1990; Zwicker and Henning, 1991; Verhey and Heise, 2012). All studies used a matching procedure to measure the level of the signal in a masking release condition that is required to produce the same partial loudness than the signal in a baseline condition where the cues eliciting a masking release are not present. Most of the studies were on the effect of BMLD on suprathreshold perception. One study also investigated how masker (co-)modulation affects suprathreshold perception of a target sinusoid (Verhey and Heise, 2012). The studies have in common that they determine the levels at equal loudness for the masking release condition and a baseline condition where the cue eliciting the masking release is not present. The studies indicated that (i) the effect of masking release decrease with level and (ii) that, at levels well above threshold, perception was hardly affected by the masking release, i.e., the partial loudness of the signal in the two masking conditions is equal at the same signal level in dB sound pressure level (SPL). Most studies concluded from their data that a masking release was no longer observed at a level of about 20 dB sensation level for the baseline condition. An exception is the data of Zwicker and Henning (1991), who found that, at low signal frequencies, an effect of binaural cues was at least measurable up to 30 dB.

One limitation of the majority of the studies on the effect of masking release on suprathreshold perception is that they show data from two to four subjects only (Townsend and Goldstein, 1972; Soderquist and Schilling, 1990; Zwicker and Henning, 1991). In the study of Verhey and Heise (2012), up to ten subjects participated. However, this study only measured sensation levels of up to 18 dB for the baseline condition, i.e., the smallest range of levels of all the above mentioned studies.

The limited number of subjects and/or signal levels above threshold is presumably due to the matching procedure used in those studies, which is a relatively time-consuming experimental paradigm. A further limitation of the matching procedure is that it does not provide direct information about the shape of the loudness function. The present study investigated if categorical scaling as a fast procedure to measure loudness functions could be used to assess the effect of masking release on suprathreshold perception of the target signal.

Partial loudness of the target sinusoid was measured for the two masking conditions of Verhey and Heise (2012) with (i) a categorical loudness scaling method and, for comparison, (ii) with a matching procedure as in Verhey and Heise (2012). In addition, categorical loudness of the sinusoid was also measured for an unmodulated masker that was reduced in level by the masking release due to masker modulation. This condition was included to test the hypothesis of Verhey and Heise (2012) that the partial loudness of a masked sinusoid in a condition of co-modulation masking release is the same as the partial loudness of the masked sinusoid for a masker with a level that is reduced by the masking release.

5.2 Methods

5.2.1 Apparatus and Stimuli

A standard personal computer controlled stimulus generation and presentation and recorded results using Matlab based software packages. The stimuli were generated digitally at a sampling rate of 44.1 kHz, D/A converted, and amplified by a Fireface 400. The participants were seated in a sound-insulated booth and the sound was presented diotically and presented via Sennheiser HD 650 headphones. The target signal was a 986-Hz pure tone. The signal duration was 600 ms including 50-ms raised-cosine ramps at on- and offset. The signal was temporally centered in the masking bandlimited noise (250 to 4000 Hz). Masker duration was 700 ms including 50-ms raised-cosine ramps. The noise was generated by transforming a white-noise into the frequency domain via a Fast Fourier Transform (FFT) and setting all Fourier components outside the desired passband to zero. A subsequent inverse FFT on the complex buffer pair yielded the waveform of the bandpass-filtered noise. The masker was either an unmodulated or a modulated noise. The modulator was an irregular unipolar (0,1) smoothed square wave as used in Verhey and Ernst (2009), Ernst *et al.* (2010), Verhey and Heise (2012), and Buss *et al.* (2012). The average duty cycle of the regular square wave was 50%, i.e., for each 25-ms period, the signal was on average switched on for half of the time. Thus the average modulation rate was 40 Hz. In order to generate irregular square waves, onset and offset times of each period were slightly jittered. The magnitude of the jitter was 10% of a period. On- and offset times were jittered independently. This jitter introduced random fluctuations of the duty cycle in the range from 30 to 70%. The square-wave modulator was smoothed by convolving it with a 5-ms raised-cosine window. For each presentation of the masker a new noise sample was used and, in the case of the modulated masker, a new modulator sample was generated. Examples of the masker waveforms (in grey) for the two noise types with an added signal (in black) are shown in Figure 5.1.

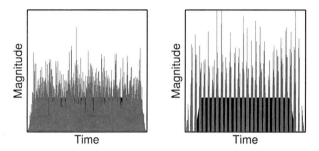

Figure 5.1: Magnitude signals of the masker (grey) and the target signal (black) for the two masker types used in the present study: unmodulated (left panel) and the square-wave modulated noise (right panel). In this example, masker and signal had the same level (55 dB).

5.2.2 Procedure

At the beginning of the experiment, masked thresholds of the sinusoidal target signal were measured for (i) the modulated masker at a masker level of 55 dB sound pressure level (SPL), referred to as the "mod condition" in the following, (ii) the unmodulated masker at the same level ("unmod condition"), and (iii) the "unmod⁻ condition" where an unmodulated masker with a reduced masker level was used. The level reduction was determined individually as the difference between the threshold in the unmod and the mod condition, i.e., the masking release. A three-alternative, forced-choice (3-AFC) procedure with adaptive signal-level adjustment (1-up 2-down) was used to determine the thresholds (Levitt, 1971). Each interval contained the masker and the sinusoidal target signal was added to one of the three intervals which was randomly selected for each trial. Intervals in a trial were separated by 500-ms silence intervals. Subjects had to indicate which of the intervals contained the target signal. Visual feedback was provided after each response. The signal level was initially adjusted in steps of 8 dB. This step size was halved after each upper reversal until a minimum step size of 1 dB was reached. The run terminated after a further 6 reversals. The average of these 6 last reversals was taken as an estimate of the threshold. The procedure was repeated three times for each condition and the average of these estimates was taken as the final estimate of the threshold.

The perception of the masked target signal at suprathreshold levels was measured using a loudness-matching procedure and categorical loudness scaling. The loudness-matching procedure was essentially the same as used in Verhey and Heise (2012). The target level at equal loudness was determined using an adaptive two-alternative, forced-choice procedure with an one-up one-down rule. Each trial consisted of one interval with a sinusoid embedded in an unmodulated noise and one interval with a sinusoid embedded in a modulated noise. Both masking noises had a level of 55 dB SPL. Note that this masker level was 10 dB lower than in Verhey and Heise (2012). The order of the intervals was randomized for each trial. Only the signal level of the test interval was varied within an adaptive track. This was either the level of the sinusoid masked by the unmodulated noise or the level of the sinusoid masked by the modulated noise. The level of the sinusoidal signal in the other (reference) interval (i.e. the interval with the fixed- level sinusoid) was 5, 10, 15, 20 or 25 dB above individual masked threshold of the respective masking condition. After each presentation of the two intervals of the trial, subjects were asked in which of the intervals the signal was louder. The step size of the adaptive procedure was initially 8 dB, reduced to 4 dB after the first upper reversal and to 2 dB after the second upper reversal. The track was continued with the smallest step size for another 4 reversals. The average of these 4 last reversals was taken as an estimate of the target level at equal loudness. To reduce potential bias effects, all ten adaptive tracks (5 reference levels x 2 masking conditions) were interleaved. This procedure was repeated three times and the average over the levels obtained in each run was taken as the final estimate of the target signal level at equal loudness. If the level of the sinusoid in a run was more than 5 dB below its masked threshold it was not used for the final averaging.

Partial loudness of the masked signal was also measured with an adaptive loudness scaling procedure as described in Brand and Hohmann (2002). The eleven-category scale was the same as used in the example of ISO 16832 (2006). The procedure consists of two phases. In the first phase, the individual dynamic range for the masked tone is determined. The levels of the following second phase are uniformly distributed over the individual dynamic range which has been estimated in the first phase. The named loudness categories in German were "unhörbar" (inaudible), "sehr leise" (very soft), "leise" (soft), "mittel" (medium), "laut" (loud), "sehr laut" (very loud), and "extrem laut" (extremely loud). Additionally, four unnamed intermediate response alternatives were represented by horizontal bars between very soft and very loud, resulting in a total of 11 categories. The same eleven-category scale was used in the example of ISO 16832 (2006). In order to avoid context effects which are due to the tendency of some listeners to rate the current stimulus relatively to the previous stimulus, the stimuli were presented in pseudo-random order where the maximum difference of subsequent presentation levels was smaller than half of the dynamic range of the sequence.

For each of the three masking conditions (mod, unmod, and unmod$^-$) the data were obtained separately, i.e., the tracks were not interleaved. In total, each subject did three categorical scaling experiments (consisting of one measurement for each masker condition) and three matching experiments. For half of the set of subjects the order was "matching – scaling – matching – scaling – matching – scaling" and for the other half it was "scaling – matching – scaling – matching – scaling – matching". For the derivation of the loudness functions, the categories were linearly transformed to numerical values (categorical units, CU) from 0 (inaudible) to 50 (extremely loud). A model loudness function was fitted to the individual data of one repetition of the procedure as described in Brand and Hohmann (2002) using a modified least-square fit. The function consists of two linear parts with independent slope values. The transition region between these linear parts was smoothed using a Bezier fit. The fit has three fitting parameters: the slope of the lower linear part, the slope of the upper linear and the level of the intersection of these two linear functions. Individual fitting parameters for each type of noise masker were estimated by calculating the median across the fitting parameters of the three runs. These median parameters were then used to determine the median loudness functions. This procedure follows the recommendations for the determination of an average loudness function in section 5.1 of the ISO 16832 (2006).

5.2.3 Subjects

Eighteen subjects (14 female, 4 male) participated in the experiments. The age ranged from 18 to 33 years (21 years on average). All subjects showed a normal audiogram in the relevant frequency range, i.e., thresholds were 15 dB HL or lower for all audiometric frequencies between 250 and 4000 Hz.

5.3 Results

5.3.1 Categorical loudness scaling

Figure 5.2 shows the individual loudness functions and the individual thresholds (stars) of the sinusoid in either of the three noise masker. Each panel shows data of one subject for a level range typically used in the previous matching studies, i.e., from the minimum threshold (for the mod condition) up to about 35 dB above the average threshold in the baseline condition. For the mod condition (filled black stars), masked thresholds obtained with the AFC procedure range from 22.0 dB SPL (subject S15) to 31 dB SPL (subject S8). On average this leads to a threshold of 27 dB SPL. For the unmod condition (oben stars), thresholds range from 33 dB SPL (subject S9) to 38 dB SPL (subject S12) with an average of 35 dB SPL. As a result, the masking release at threshold, e.g., the level difference between the unmod and the mod thresholds, ranged from 3 dB (S8) to 12 dB (S6) with an average of 8 dB. For the unmod$^-$ condition (smaller filled grey stars), the lowest threshold (24 dB SPL) was found in subject S6, the highest (35 dB SPL) in subject S8. The average threshold for this condition was 29 dB SPL. For all subjects, the lowest threshold was observed for the mod condition, and the highest for the unmod condition. The threshold for the unmod$^-$ condition lay between those of the other two. In general it was closer to the threshold in the mod condition as intended (12 out of 18 subjects). However, for three subjects (S16, S17 and S18) it was closer to the threshold for the unmod condition, for subjects S5 and S11 threshold in the unmod$^-$ condition was the lowest threshold and for subject S8 it was the highest threshold.

For most subjects, thresholds measured with the AFC procedure were underestimated by the level of the loudness function at the named category inaudible (0 CU) of the loudness function. Two subjects (S1 and S14) had lower thresholds estimated from the scaling data for the unmod (black dashed line) and the mod (straight black line) condition as for the unmod$^-$ condition (dashed grey line). However, for some subjects (e.g., S9) the relation of the thresholds was within the same range as found in the AFC measurement. All individual and average thresholds are shown in Table 5.1.

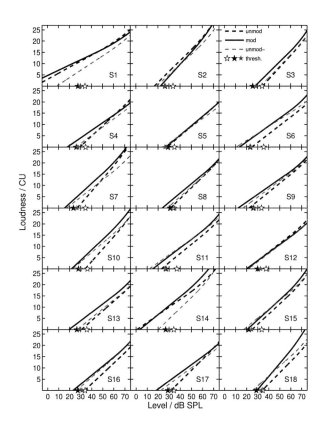

Figure 5.2: Individual loudness growth functions for the eighteen subjects participating in the experiments. Each panel shows data of one subject. Different line styles indicate different masking conditions: mod (solid line), unmod (black dashed line), and unmod$^-$ (grey dashed line).

The black stars indicate the thresholds for the mod
(filled star) and the unmod (open star) conditions, the
grey star indicates the unmod$^-$ condition.

For most subjects, thresholds measured with the AFC procedure were underestimated by the level of the loudness function
at the named category inaudible (0 CU) of the loudness function. Two subjects (S1 and S14) had lower thresholds estimated
from the scaling data for the unmod (black dashed line) and the
mod (straight black line) condition as for the unmod$^-$ condition
(dashed grey line). However, for some subjects (e.g., S9) the relation of the thresholds was within the same range as found in
the AFC measurement. All individual and average thresholds
are shown in Table 5.1.

As expected, loudness increased with increasing level for all subjects and conditions. In general, the scaling data indicate that
subjects with low masking release at threshold (e.g., S8 and
S2) also had a low masking release at suprathreshold levels.
No difference between the loudness functions for the three conditions was found for subjects S5 and S12. Small differences
in slope and magnitude were found for subjects S2 and S15.
Subjects S1 and S14 showed similar loudness functions for the
unmod and the mod condition. The slope for the unmod$^-$ condition was higher, resulting in a decreasing level difference between the curves with increasing level. Although the individual
magnitudes differed, subjects S6, S9, S11, S16 and S17 showed
similar patterns for the different loudness functions. These subjects had a steeper loudness function for the unmod condition
as for the other two conditions. The loudness functions of these
latter conditions differed only slightly in magnitude and slope.
Comparable results were found for S7 and S10, only they showed

an increasing difference between the unmod⁻ and the mod condition with increasing signal level.

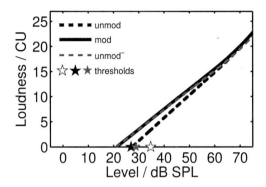

Figure 5.3: Average loudness growth functions for the eighteen subjects participating in the experiments. Different line styles indicate different masking conditions: mod (solid line), unmod (black dashed line), and unmod⁻ (grey dashed line). The black stars indicate the thresholds for the mod (filled star) and the unmod (open star) conditions, the grey star indicates the unmod⁻ condition.

The trends described in the majority of the individual data were also observed in the average data, as shown in Fig. 5.3. The data representation is the same as for the individual data in Fig. 5.2. The loudness function in the unmod condition has a steeper slope than the loudness functions of the mod and the unmod⁻ conditions. The latter two functions were about equal in the depicted level range. For signal levels higher than 60 dB SPL, equal levels in dB SPL elicited the same loudness in all masking conditions. As for the individual data, the average loudness functions also underestimated the AFC-thresholds by

5 to 8 dB in all conditions. However, general relations between thresholds were comparable for the two measurement procedures. The threshold for the mod condition was lower than that for the unmod condition. The thresholds for the mod and the unmod$^-$ conditions were about equal when estimated by the loudness function. When measured with the AFC-procedure, the unmod$^-$ masker lead to an about 2 dB higher threshold.

5.3.2 Loudness matching

Figure 5.4 shows individual matching data for the 18 subjects. Each panel shows the data in dB SPL of one subject. The level range is the same as the one used in the previous two figures showing the scaling data. The level of the sinusoid in the mod condition is plotted against the level of the sinusoid in the unmod condition. Open circles indicate data points from the adaptive track where the signal masked by the unmodulated noise was varied and closed circles those where the level of the signal masked by the modulated noise was varied. Error bars denote standard deviations across the three repetitions. The star indicates the threshold for the two masking conditions. For all subjects, the star lies below the diagonal indicating that the threshold for the mod was lower than that for the unmod condition. This release of masking can also be seen at suprathreshold levels up to 15 (subject S17) to 25 dB (e.g., subject S7) above the threshold in the unmod condition. In general, subjects show similar patterns for the suprathreshold perception of the sinusoid. For low to moderate levels above threshold, all data points are below the diagonal. At high levels, data of about half of the subjects lie on the diagonal, for two of them sometimes even above the diagonal (S2 and S4). For the others the level of the sinusoid in the mod condition was below that of the sinusoid in the unmod condition. This general trend is also confirmed by the linear regression line fitted to the individual data (thin

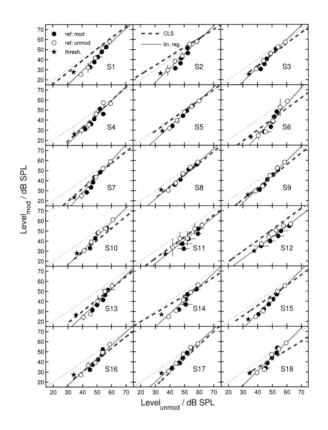

Figure 5.4: Individual data for the partial loudness of the sinusoid in the mod and the unmod condition (open and closed circles), respectively. The level of the sinusoid in the mod condition is plotted over the level of the sinusoid in the unmod condition. In addition, a linear regression

through the data points including masked thresholds (black star) is shown by the thin black line. For later comparison, the dashed grey line indicates the results of the scaling procedure. Each panel shows data of one subject and the error bars denote the individual standard deviations.

solid black line in Fig. 5.4. Differences occur in the exact shape of the individual curves, e.g., in steepness and magnitude. Subjects S11 and S12 show an approximately linear increase parallel to the diagonal, whereas others (e.g., subjects S3 and S4) show a linear increase with a slope higher than 1. Note that most of the subjects show a slightly non-linear increase with increasing level to a greater or lesser extent which is not captured by the linear regression function. A very pronounced non-linearity is found for subject S7 with hardly any increase at low levels above threshold, a steep slope at medium levels and an increase about parallel to the diagonal at high levels.

Figure 5.5 shows average matching data across the 18 subjects. As in Figure 5.4, the left panel shows the level of the target signal in the mod condition over the level of the target signal in the unmmod condition. Symbols are the same as used in Figure 5.4. In general, the data points for equal loudness are below the diagonal, i.e., the sound pressure level of the tone signal in the mod condition is smaller than that of the tone signal in the unmod condition. However, this level difference at equal loudness decreases with increasing level, showing that the advantage due to the lower threshold in the presence of the modulated masker decreases with the increasing sensation level of the signal. This finding is again confirmed by the linear regression line. At the highest level of about 25 dB above the threshold in the unmod condition hardly any masking release is observed in the average data. Thus, the same level of the tone in the two masking conditions elicits about the same sensation of loudness. The right panel shows the same set of data but expressed in dB relative to the masked threshold. For this representation, data point are slightly below the diagonal for levels up to about 15 dB. For higher levels, data points are above thresholds, indicating that the level of the tone at high levels above threshold is higher in the mod than in the unmod condition. Thus the non-linearity found in most individual data is also observed in the average data.

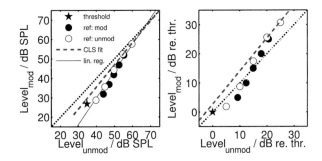

Figure 5.5: Average data for the partial loudness of the sinusoid in the mod and the unmod noise across indiviual data for each reference condition seperately. The symbols and the data representation in the left panel is the same as in Figure 5.4. The right panel shows the same data in dB relative to the masked threshold.

5.4 Discussion

5.4.1 Comparison of loudness matching and loudness scaling data

The dashed grey lines in Figures 5.4 and 5.5 indicates levels at equal loudness derived from the scaling data shown in Figs. 5.2 and 5.3. A very good match of individual results of the two procedures (i.e., the scaling data and the linear regression) is found for subjects S4, S8, S16 and S17. The differences in slope were less than 0.1 and the differences in y-intercept were less than 5 dB except for subject S16 with differences of 0.16 and about 7 dB, respectively. Low differences in slope were also found for S11 and S12. However, for them the graphs for scaling and matching data were shifted against each other, leading to differences in y-intercept of about 7 to 8 dB. The general trend of the matching data is also seen in the scaling data for some other subjects (e.g., S7 and S13) but with differences in slope of up to 0.35 and up to about 14 dB in y-intercept. For subjects S2, S14 and S18, loudness matching data show a steep slope, whereas the slope of the scaling data for these subjects is lower than 1. The greatest differences in slope (0.56) and y-intercept (32 dB) were found for subjects S2 and S18, respectively. Individual and average slopes and y-intercepts for both procedures are listed in Table 5.1.

A comparison of the average results of the two procedures shows that the effect of masking release is less pronounced in the results of the loudness scaling procedure than in the matching data. The scaling data capture the general trend of the matching data revealing an effect of masking release up to levels of the sinusoid in the unmod condition of about 60 to 65 dB SPL. Thus, for both measurement procedures the effect of release from masking is measurable up to about equal absolute levels in dB SPL of the sinusoid. Note that thresholds of the sinu-

Table 5.1: Comparison of matching (Match) and scaling (CLS) data (columns 2-5) and comparison of thresholds (Thr.) measured with the AFC procedure and those derived from the CLS data (columns 6 - 11). The first comparison is done on the basis of the slopes and y-intercepts of the linear regressions for the matching data and the equal-loudness curves derived from the CLS data (thin black line and dashed grey line in Figs. 5.4 and 5.5). The average values shown in the bottom row are derived from the average data.

Subject	(lower) slope Match	(lower) slope CLS	Y-intercept [dB] Match	Y-intercept [dB] CLS	Thr. unmod [dB] AFC	Thr. unmod [dB] CLS	Thr. mod [dB] AFC	Thr. mod [dB] CLS	Thr. unmod- [dB] CLS
S1	1.39	1.15	-26.6	-7.2	34.0	-11.1	27.6	-20.0	8.6
S2	1.40	0.90	-24.9	7.3	33.7	16.8	26.4	22.3	18.9
S3	1.40	1.11	-24.3	-10.6	34.1	34.7	25.6	27.8	32.9
S4	1.42	1.34	-26.3	-21.3	34.5	29.4	26.1	18.2	25.8
S5	1.31	1.02	-18.9	-1.8	34.9	28.6	29.2	27.5	25.5
S6	1.60	1.08	-40.1	-13.1	35.5	25.0	23.2	13.8	10.8
S7	1.59	1.26	-35.9	-18.3	34.3	26.9	23.4	15.6	17.5
S8	1.24	1.16	-15.7	-12.2	34.0	31.3	30.9	24.0	28.6
S9	1.40	1.16	-24.2	-13.4	33.4	22.4	25.9	12.5	16.9
S10	1.41	1.19	-26.1	-18.7	36.1	34.2	27.9	21.9	22.9
S11	1.16	1.08	-17.2	-9.7	36.9	23.3	27.2	15.4	12.3
S12	1.08	1.07	-10.1	-2.1	37.6	20.8	29.9	20.1	21.9
S13	1.37	1.15	-29.6	-15.9	35.5	30.9	26.5	19.7	27.2
S14	1.33	0.86	-25.5	2.7	34.2	1.1	26.9	3.6	20.3
S15	1.46	1.02	-30.9	-4.9	33.6	25.1	22.0	20.6	17.5
S16	1.31	1.15	-22.2	-15.6	33.9	33.8	27.3	23.2	26.3
S17	1.32	1.36	-19.3	-25.0	34.2	31.9	28.8	18.4	22.6
S18	1.41	0.85	-25.7	0.2	34.5	34.2	28.7	29.4	25.5
average	1.37	1.13	-24.8	-9.8	34.7	27.5	26.9	21.4	21.3

soid in the mud and the unmod condition are about 8 dB lower when derived from the scaling data. So only based on scaling data, a masking release seems to be observed up to about 33 dB above unmod threshold instead of about 25 dB above unmasked threshold based on the AFC data. The comparison of the average slopes and the y-intercepts reveals differences of about 0.23 and 15 dB, respectively, indicating that at least general trends of the matching data are reproduced by the scaling procedure. An advantage of the scaling procedure is the overall measurement time. One single scaling run lasts about two minutes at the maximum. This results in a total measurement time of about 15 to 20 minutes for the nine scaling runs in the present experiments. On the contrary, one measurement block of the matching procedure consisting of 10 tracks already takes 15 to 20 minutes. Thus, the overall duration of three matching blocks adds up to 45 to 60 minutes.

5.4.2 Comparison with literature data on masked thresholds

When the masked thresholds are expressed relative to the masker level (i.e., level in dB SPL minus 55 dB), the average thresholds of the present study are about -20 dB for the unmod condition and -28 dB for the mod condition. The unmod threshold is slightly lower than the -17 dB in Ernst et al. (2010) but comparable to the -19 dB in Verhey and Heise (2012). Both studies used a slightly higher masker level but several studies have shown that at these masker levels threshold for an unmod condition increase more or less linearly, i.e., by about 10 dB per 10-dB increase of masker level (e.g., McFadden, 1968; Hall et al., 1984; Moore and Shailer, 1990; Nitschmann et al., 2009). This is also in agreement with the present study.

In contrast, thresholds in masking release conditions increase slightly less with level resulting in an increase in masking release as the level is increased, even at the moderate masker level used here (e.g., Moore and Shailer, 1990, for CMR). This may partly account for the smaller masking release in the present study (8 dB) compared to 15-16 dB in Verhey and Heise (2012) and Buss *et al.* (2012) who both used a 10 dB higher masker level. Another difference between these two studies and the present study is the amount of jitter. Verhey and Heise (2012) and Buss *et al.* (2012) used a jitter of 20% whereas the jitter was only 10% in the present study. Verhey and Ernst (2009) showed that the masking release decreased by about 2 dB when the jitter was decreased from 20 to 10%. However, this jitter difference is presumably not the main factor, since Ernst *et al.* (2010) measured a masking release of 15 dB for a 62-dB masker level with the same jitter as used in the present study. Thus it is likely that the main differences between the studies are the groups of subjects participating in the studies. Verhey and Ernst (2009) and Verhey and Heise (2012) both reported large individual differences for the masking release with square modulations. For 10% jitter, the masking release ranged from about 8 dB to 20 dB. The individual variation in the present study was slightly smaller but in contrast to the previous studies, some subjects had very small masking release down to only 3 dB (subject S8). However, two third of the subjects of the present study had masking releases of 8 dB or higher, i.e., they had masking releases in the range of those found in Verhey and Ernst (2009).

5.4.3 Comparison with literature data at suprathreshold levels

The matching data indicate hardly any masking release at 25 dB above the threshold in the baseline (unmod) condition. This is in agreement with most of the previous studies comparing the loudness of a sinusoid in a masking release condition to that of a sinusoid in a baseline condition (Townsend and Goldstein, 1972; Soderquist and Schilling, 1990; Verhey and Heise, 2012). Zwicker and Henning (1991) showed that, at one frequency (250 Hz), a residual masking release due to binaural cues can be observed up to levels of 30-40 dB. The categorical scaling data of the present data indicate such a residual masking release at levels above 25 dB also for a CMR paradigm at 986 Hz. The present matching data indicate a shallower slope for levels just above thresholds followed by a steeper portion for the next 10 to 20 dB before asymptotically reaching the diagonal at very high levels. A similar behavior is observed in Verhey and Heise (2012). For comodulated maskers it was shown that the psychometric function tends to be slightly steeper for the mod condition than for a baseline condition (Buus *et al.*, 1996; Bacon *et al.*, 2002).Thus the increase in audibility with signal level is faster for the mod condition than for the unmod condition which may explain the smaller increase in level at equal loudness for the mod condition than for the unmod condition for levels slightly above threshold.

Whereas the matching experiment only investigate a change in perception between two masking release conditions the scaling experiment provides information of the loudness growth function. Figure 5.6 shows the scaling data transformed into sones using the equation relating CU to sone of Heeren *et al.* (2013). In addition, the standardized loudness growth function (dotted line) of a 1000-Hz sinusoid is shown (ANSI S3.4, 2007). The present data confirm previous findings on the loudness of masked signals. They found that not only the threshold was elevated due to the presence of the masker but that also the loudness was affected at suprathreshold levels (Lochner and Burger, 1961; Gleiss and Zwicker, 1964). The slope of the loudness function was higher than that of the unmasked signal at levels close to threshold whereas the loudness functions were similar at very high levels.

In general, the loudness functions for the mod (solid black line) and the unmod⁻ condition (dashed grey line) are about the same. Only at levels above about 85 dB SPL, the loudness function of the mod condition becomes slightly steeper. The

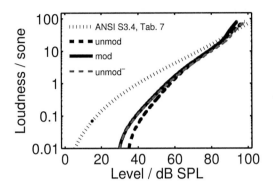

Figure 5.6: Average loudness growth functions in sones for the 18 subjects participating in the experiments. As in Fig. 3, results of the mod condition are shown with a solid line, those of the unmod condition with a black dashed line and of the unmod⁻ conditions with a grey dashed line. The CU value were transformed sones using the transformation on the basis of Heeren *et al.* (2013). In addition to the experimental data, the standardized loudness growth function of an unmasked 1000-Hz sinusoid with a dotted line (ANSI S3.4, 2007).

very good agreement over a large loudness range supports the hypothesis of Verhey and Heise (2012) that the loudness function for a condition of masking release is similar to a masking condition without a masking release but with a reduced masker level.

Note that at very high loudness the measured loudness functions of all masking conditions differ slightly from the standardized loudness function of ANSI S3.4 (2007) for an unmasked signal. This may indicate limitations of the fitting procedure since Lochner and Burger (1961) found that the loudness functions of masked signals equal the loudness functions of unmasked signals at very high levels. Such limitations of the categorical scaling procedure were, e.g., reported in Al-Salim *et al.* (2010). They found that the slopes of the loudness functions for lower levels were essentially the same in a test-retest comparison whereas the slopes for higher levels were not reliable across the two sessions (however, see Rasmussen *et al.*, 1998). A higher inaccuracy of the procedure at high levels may be due to a lower number of stimuli presented at high levels than at low levels (Brand and Hohmann, 2002).

5.4.4 Relation to neural correlates of masking release

Several studies in the literature investigated neural correlates of masking release using electroencephalography (EEG, e.g., Fowler and Mikami, 1996; Wong and Stapells, 2004; Androulidakis and Jones, 2006; Ishida and Stapells, 2009; Epp *et al.*, 2013) magnetoencephalography (MEG,e.g., Sasaki *et al.*, 2005; Rupp *et al.*, 2007; Ross *et al.*, 2007) and functional magnetic resonance imaging (fMRI Ernst *et al.*, 2010; Wack *et al.*, 2012). Since levels in these studies are usually above threshold it would have been interesting to also measure psychophysically how the sensation changes with level at such suprathreshold levels. The data which were presented in those studies look promising also when analyzed with respect to partial loudness at a suprathreshold level. For example, Sasaki *et al.* (2005) found for a 250-Hz sinusoidal signal embedded in a diotic masker that the N1m, i.e., the first prominent peak in the MEG response at 80 ms after the onset of the signal, increases more rapidly for an antiphasic signal than for a diotic signal (their panel A of Fig. 2). At a level of about 20 dB above the diotic threshold the suprathreshold binaural unmasking effect disappeared. This is in agreement with the psychophysical results of Soderquist and Schilling (1990) who, at this frequency, also did not show unmasking at 20 dB above the diotic condition (however, see Zwicker and Henning, 1991).

In the second analysis of Ernst *et al.* (2010), they specifically focused their attention only on regions where the response to the modulated masker was smaller than that for the unmodulated masker. The assumption of such a lower activation for a modulated masker than for an unmodulated masker for a cortical region showing a masking-release is in agreement with the current finding of a similar loudness function of the masked signal for the mod and unmod$^-$ condition. The cortical regions fulfilling this requirement not only showed a higher sensitivity to the signal in the mod condition than in the unmod condition but also showed that the increase with signal level was smaller than for the signal in the unmod condition. The ratio of the slopes was 1.3. This is similar to the ratio of the slopes of about 1.4 derived from the matching data (thin black line in Fig 5.5) and 1.1 derived from the lower portion of the categorical loudness functions shown in Fig 5.3. Neither Ernst *et al.* (2010) nor Sasaki *et al.* (2005) directly measured suprathreshold perception in their groups of subjects, presumably due to time consuming matching procedure commonly used to investigate this perception. The present study indicate that at least general trends may be already quantified by using a fast scaling procedure.

5.5 Summary

In summary, two procedures were used to investigate suprathreshold perception of a masked signal under a condition of masking release due to masker amplitude modulation. Although a matching experiment as used in previous studies on suprathreshold effects of a cue eliciting a masking release provide a detailed view on this perception a categorical scaling procedure is useful to investigate general trends. Such scaling may allow to investigates the neural correlates of suprathreshold masking release without a substantial increase in measurement time.

Which of the two procedures (matching or scaling) is going to be used in psychophysical studies depends on the required information of the results. The scaling procedure was used in the present study to test the hypothesis that suprathreshold perception in a masking release condition is similar to the perception of a signal embedded in a noise which is reduced in level by the masking release at threshold. The scaling data confirmed this hypothesis.

6 Categorical scaling of partial loudness in CI users

6.1 Introduction

In our daily life we are surrounded by sounds which we have to classify as those containing relevant messages such as speech acts or meaningful noise in a certain situation, e.g. an approaching car versus those sounds which are non-relevant in a specific context. The latter often vary in direction, amplitude, frequency or even familiarity and tend to interfere with sounds containing relevant information. In order to identify, separate and process acoustic inputs, both monaural and binaural cues are of great use. Additionally, auditory attention and memory play an important role for the information identification. The acoustic cues, e.g. depend on the location of the sound in space. For instance, time (or phase) and level differences appear at the two ears due to difference distances to the source or head-shaddow effects. These spatial cues can help to improve speech understanding in noise in situations where signal and noise are spatially separated compared to listening situations where both signals originate from the same direction (see, e.g., Plomp and Mimpen, 1981; Hawley *et al.*, 1999; Litovsky, 2005). This benefit due to spatial constellations is referred to as spatial release from masking (SRM Litovsky *et al.*, 2012). In humans

SRM is usually studied by investigating the speech understanding in noise either by determining differences in signal-to-noise ratio or speech reception threshold for different spatial combinations of speech and masking signals. Different monaural as well as binaural effects such as the "better ear effect" (also referred to as "monaural head shaddow effect", van Wanrooij and van Opstal, 2004), "binaural squelch", or "binaural summation" can affect the SRM (Nopp *et al.*, 2004). Determining the SRM by means of signal-to-noise ratios and speech reception thresholds already include rather complex signals. A more basic method to determine the SRM is the comparison of detection thresholds of a tone masked by noise for different conditions. The differences in threshold are referred to as binaural masking level difference (BMLD). Depending on the condition BMLDs of up to 30 dB can be reached. BMLDs are often measured by means of headphones and introducing interaural time (ITD) or phase differences (IPD), e.g. by van de Par and Kohlrausch (1997). Headphone measurements provide a very reliable sound field. On the other hand, they do not include other acoustic cues that are available in realistic listening conditions such as free-field measurements. Moreover, they are insufficient for listeners who are in need of different kinds of hearing devices. This group of listeners includes, e.g. cochlear implant (CI) users. CIs are the only hearing devices that can, to a certain extend, restore hearing in humans that suffer from severe-to-profound hearing loss by directly stimulating the neurons in the auditory nerve. Yet, much remains to be done in order to improve the hearing performance with CIs, especially regarding speech perception in noisy backgrounds. Rather than investigating the SRM in CI users directly, studies have been conducted that investigate SRTs (Litovsky *et al.*, 2009) or focus on the benefit of bilateral in contrast to monaural implantation (Nopp *et al.*, 2004). Some studies (e.g., Schleich *et al.*, 2004; Litovsky *et al.*, 2009) reveal, that bilateral improvement results from monaural cues. The authors suggested a benefit of bilateral over monaural usage while the performance remains worse compared to normal-hearing listeners. This suggestion is confirmed by Loizou *et al.* (2009) who investigated the SRM preserving spatial cues by means of head related transfer functions (HRTFs). Stimuli were directly presented through the auxiliary input jack of a provided research processor. The authors results reveal an improvement in speech perception for CI users of 3 to 4 dB when presenting the masking noise spatially apart from the speech signal. By contrast, in the normal-hearing control group (data taken from, Hawley *et al.*, 2004) a reduction in SRT of about 6 dB can be observed. In the same study, measurements of the SRM using only monaural cues showed equal results for both groups of listeners. In fact, for CI users monaural benefit was not considerably less than the overall benefit of 3 to 4 dB. Thus, Litovsky (2012) suggested that bilateral CIs are not effective at preserving binaural cues. Reasons for this might be not well-preserved and/ or synchronized inputs at the left and right ear.

As compared to direct stimulation in CI users or headphone measurement in normal-hearing listeners, only free-field measurements provide the usage of realistic spatial cues. Hence, the present study focuses on the basic measurement of spatial release from masking in a free-field condition by investigating the threshold and suprathreshold perception of a sinusoid masked by a Gaussian noise that is spectrally centered around the sinusoid. For this purpose loudness perception of the tone over the whole auditory dynamic range is measured for different spatial positions of the masking noise using categorical loudness scaling. By means of this procedure, BMLDs are determined for cochlear implant users as well as for a normal-hearing control group.

6.2 Methods

6.2.1 Listeners

Seven bilaterally implanted CI users (3 male, 4 female) participated in the experiment. Listeners CI1, CI3, CI4, CI5, CI6, and CI7, used MED-El® CIs (models C40+, Pulsar, Sonata, Concerto) and were fitted with OPUS 2 sound processors. CI 2 (45 years, male) was implanted on both sides contemporary with Nucleus® CIs (model CI24RE) and provided with CP910 sound processors. Detailed listener demographics are shown in Table 6.1.

The normal-hearing control groups consisted of ten adults (5 males, 5 females) at ages between 21 and 35 years. All had hearing thresholds ≤15 dB HL at standard audiometric frequencies between 250 and 4000 Hz.

6.2.2 Apparatus and Stimuli

Stimulus generation and presentation and the recording of the results was controlled by a standard personal computer and Matlab based software packages. All stimuli were generated at a sampling frequency of 44.1 kHz, D/A converted using a RME ADI-8 QS and presented via three Tannoy Reveal 5A loudspeakers. The speakers were positioned at a distance of 1.5 from the listener at angles of -60, 0 and +60° on the horizontal plane. The target signal was a 500 Hz pure tone. The signal duration was 1000 ms including 50-ms raised-cosine ramps at signal on- and offset. The signal was masked by a Gaussian one-octave wide bandpass noise with a center frequency of 500 Hz. The masker duration was also 1000 ms including 10-ms raised-cosine ramps at on- and offset.

Table 6.1: CI user demographics

CI user	Sex	Ear	Age at implant	Device	Processor	Strategy	Duration of use (years)
CI1	m	R	47	Concerto	OPUS 2	FS4	3.5
		L	48	Concerto	OPUS 2	FS4	2.5
CI2	m	L	44	CI24RE	CP910	ACE	1
		R	44	CI24RE	CP910	ACE	1
CI3	f	L	14	Sonata	OPUS 2	FS4	13
		R	20	Pulsar	OPUS 2	FS4	7.5
CI4	f	R	1	Sonata	OPUS 2	FSP	14
		L	4	C40+	OPUS 2	FSP	11
CI5	m	R	49	C40+	OPUS 2	FSP	12
		L	52	Pulsar	OPUS 2	FSP	8.5
CI6	f	L	1	Sonata	OPUS 2	FS4	14
		R	4	C40+	OPUS 2	FSP	11
CI7	f	R	49	Sonata	OPUS 2	FSP	7
		L	51	Sonata	OPUS 2	FSP	5

6.2.3 Procedure

The loudness of the pure tone was measured binaurally in free-field for four different signal conditions. In all conditions the pure tone was presented from the front. In the first condition (S_0) no masking noise was added to the tone. In the remaining four conditions an additional masking noise was presented from the left (S_0N_{-60}), the frontal (S_0N_0), and the right speaker (S_0N_{+60}), respectively. The procedure used to determine the suprathreshold loudness perception was an adaptive loudness scaling procedure according to Brand and Hohmann (2002) using an eleven-category scale as in the example of ISO 16832 (2006). The named loudness categories in German were "unhör-bar" (inaudible), "sehr leise" (very soft), "leise" (soft), "mittel" (medium), "laut" (loud), "sehr laut" (very loud), and "extrem laut" (extremely loud). Additionally, four unnamed intermediate response alternatives were represented by horizontal bars between very soft and very loud, resulting in a total of 11 categories. The procedure consists of two phases. In the first phase, the individual dynamic range for the according stimulus condition was determined. The levels of the following second phase were uniformly distributed over this range. In order to avoid context effects which are due to the tendency of some listeners to rate the current stimulus relatively to the previous stimulus, the stimuli were presented in pseudo-random order where the maximum difference of subsequent presentation levels was smaller than half of the dynamic range of the sequence.

For each of the four signal conditions (S_0), S_0N_{-60}, S_0N_0, and S_0N_{+60}), the data were obtained separately, i.e. the tracks were not interleaved. In total, each normal-hearing listener did twelve categorical scaling runs (3 repetitions × 4 conditions), whereas each CI user performed eight runs in total (2 repetitions × 4 conditions), except for listener CI4 who only measured each condition once.

The loudness functions were derived by linearly transforming the categories into numerical values (categorical units, CU) from 0 (inaudible) to 50 (extremely loud). A model loudness function was fitted to the individual data of one repetition of the procedure as described in Brand and Hohmann (2002) using a modified least-square fit. The function consists of two linear parts with independent slope values. The transition region between these linear parts was smoothed using a Bezier fit. The fit has three fitting parameters: the slope of the lower linear part, the slope of the upper linear and the level of the intersection of these two linear functions. Individual fitting parameters for each condition were estimated by calculating the median across the fitting parameters of the respective number of runs. For the normal-hearing control group, an overall median loudness function for each condition was determined by calculating the median over the individual fitting parameters. This procedure follows the recommendations for the determination of an average loudness function in section 5.1 of the ISO 16832 (2006).

6.3 Results

Normal-hearing listeners

Figure 6.1 shows a typical result of a normal-hearing listener (NH8). Each panel shows data of one measurement condition. Different markers indicate the answers given by the listener in the three different runs. The corresponding fitting functions are indicated using different lines. In each panel, the median loudness function for the respective condition is shown by a solid line. Compared to the other three conditions, condition S_0 leads to a flatter loudness function with an estimated threshold of 1 dB. The median loudness functions for the remaining three conditions resembling each other in slope and magnitude. For the $S_0 N_0$ condition listener NH8 had a threshold of about 42 dB,

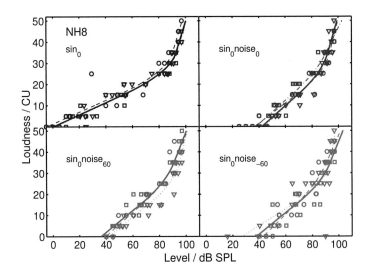

Figure 6.1: Individual results of listener NH8 as a typical example for the results of normal-hearing listeners. Each panel shows the results for one of the stimulus conditions. The different symbols in each panel denote the answers and the dashed, dotted and dashed-dotted lines the corresponding fitting functions for each run, separately. The solid line indicates the median loudness function over the three runs.

whereas for the S_0N_{60} and S_0N_{-60} conditions thresholds were about 36 and 38 dB, respectively.

Figure 6.2 shows the results of all normal-hearing listeners for the four stimulus conditions. Similar patterns of the different loudness functions are found for most listeners, although individual slopes and magnitudes differed. In general, the S_0 condition (thick straight line) resulted in the flattest loudness

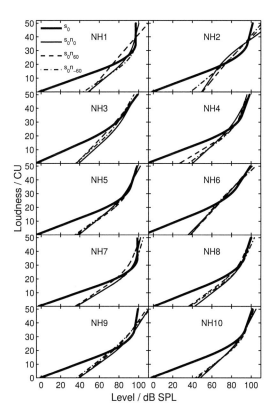

Figure 6.2: Individual results of normal-hearing listeners participating in the experiments. Each panel shows data of
one listener. Different stimulus conditions are indicated
using different lines: S_0 (thick solid line), S_0N_0 (thin
solid line), S_0N_{60} (thin dashed line), and S_0N_{-60} (thin
dash-dotted line).

function with the lowest threshold. For all listeners the lower portion of the loudness function had a shallower slope compared to the upper portion, which for some listeners (NH1 and NH7) was even vertical. Thresholds for this condition ranged from -10 dB (listener NH6) to 1 dB (listener NH8) with -4 dB on average (see also Figure 6.3). Half of the listeners (NH5, NH6, NH8, NH9 and NH10) showed about identical loudness functions for the remaining three conditions (masking conditions). The course of the functions also differed only slightly between listeners. The loudness functions of listener jg were straight lines with identical slopes for the lower and the upper portion. For listener NH9 the slopes of the upper and lower portions were also about equal. Listeners NH5, NH8 and NH10 had a slightly higher slope for the upper portions of the loudness functions. Greater differences in the loudness functions for the masking conditions were found for listeners NH1, NH2, NH3, NH4 and NH7. At least for one condition, in listeners NH1 (S_0N_{60}, dashed line) and ap (S_0N_0, thin straight line, and S_0N_{60}) the lower portion of the loudness function had a higher slope compared to the corresponding upper portion. Listener NH3 had similar thresholds and similar uncomfortable levels for the three masking conditions. However, the slopes of the loudness functions differed leading to differences for medium loudness. For listener NH4 loudness functions were about equal at medium and high loudness but differed at lower loudness, resulting in different thresholds for the three masking conditions. Listener NH7 showed a similar pattern but with differences at high loudness. Thresholds reached from 36 (NH6) to 50 dB (NH2), from 27 (NH4) to 50 dB (NH2 and NH1) and from 35 (NH5) to 44 dB (NH1) for the S_0N_0, the S_0N_{60} and the S_0N_{-60} condition (dash-dotted line), respectively.

The median loudness functions for all 10 normal-hearing listeners are shown in Figure 6.3. Data representation is the same as in Figure 6.2. Median loudness functions showed similar patterns as compared to most of the individual functions. At high

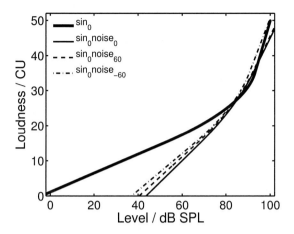

Figure 6.3: Median results of the normal-hearing control group. Data representation is the same as for the individual results in Figure 6.2. Different stimulus conditions are indicated using different lines: S_0 (thick solid line), S_0N_0 (thin solid line), S_0N_{60} (thin dashed line), and S_0N_{-60} (thin dash-dotted line).

loudness above about 30 CU similar loudness functions were found for all conditions. On the contrary, the lower portion of the loudness functions differed for the different conditions. Again, the S_0 condition revealed the loudness function with the lowest slope leading to a median threshold of about -4 dB. Steeper and about equal slopes were found for the three masking conditions. Of these, the S_0N_{-60} had the lowest slope and the lowest threshold of -37 dB. Slightly higher slopes and thresholds (41 and 44 dB) were found for the S_0N_{60} and S_0N_0, respectively.

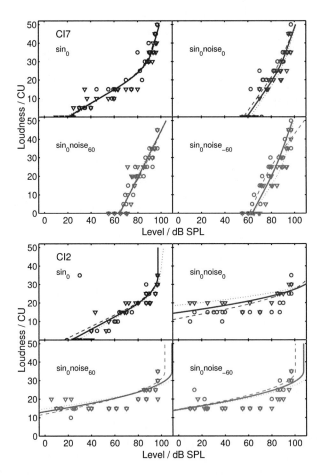

Figure 6.4: Individual results of CI user CI2 and CI7 as a typical
example for the results of CI users that could and could
not perform the task. Each panel shows the results for
one of the stimulus conditions. The different symbols in
each panel denote the answers and the dashed, dotted
and dashed-dotted lines the corresponding fitting functions for each run, separately. The solid line indicates
the median loudness function over the two runs.

CI users

On the basis of the results, CI users can be divided into two groups. In Figure 6.4 the individual results of CI users CI2 and CI7 are shown as typical examples of these two groups. The data representation is the same as in Figure 6.1. CI user CI7 represents the group of CI users that were able to rate the loudness of the sinusoid in the noise masker. Despite greater variation in the single ratings (single markers), the median loudness functions for the different conditions revealed a pattern comparable to that of the normal-hearing listeners. The S_0 condition also lead to the loudness function with the flattest lower slope, whereas the slopes of the upper portions of the loudness functions are about equal for all conditions. The lower slopes of the three masking conditions are also about equal. However, differences in contrast to the normal-hearing listeners are found when comparing the thresholds. For the unmasked condition, CI user CI7 has a threshold of about 20 dB, the masked thresholds ranged from 59 dB (S_0N_0) to 64 dB (S_0N_{60}). CI user CI2 represents the group of CI users who had difficulties to rate or even hear the sinusoid in the masking noise. Except for the S_0 condition, loudness functions considerably differ from those of the first group and the normal-hearing listeners. For CI user CI2 the lower slope and the threshold (23 dB) in the S_0 were comparable to CI user CI7. Differences for this condition, however, can be seen in the upper portion of the loudness function. The upper portion of the loudness function above 25 CU for CI user CI2 is estimated by few measurement points. Hence, the fitting function becomes vertical in that range. For the masked conditions, levels up to 80 dB are rated within the same range (between 10 and 20 dB). Between 80 and 100 dB SPL loudness values ranged between 20 and 30 CU leading to a very flat loudness function in this level range. For conditions S_0N_{60} and S_0N_{-60} this lead to one loudness function with a vertical upper portion.

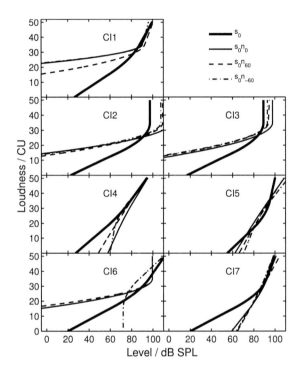

Figure 6.5: Individual results of the CI users participating in the experiment. Each panel shows data of one CI user. Data representation is the same as for the individual normal-hearing results in Figure 6.2. Different stimulus conditions are indicated using different lines: S_0 (thick solid line), S_0N_0 (thin solid line), S_0N_{60} (thin dashed line), and S_0N_{-60} (thin dash-dotted line).

Figure 6.5 shows the results of all CI users for the four stimulus conditions. CI users CI1, CI3, and CI6 show loudness functions comparable to those of CI user CI2. For the S_0 condition the shape of the loudness function of CI user CI3 is the same as for CI2, but slightly shifted to lower levels leading to an estimated threshold of 17 dB. The loudness function measured for the masking conditions were also comparable for these two CI users with the transition regions between the lower and the upper portion also shifted to lower levels for CI3. Despite a shallower slope of the upper portion of the loudness functions and a threshold of 27 dB in the S_0 condition, the results for CI user CI1 were also in the same range as for CI2 and CI3. Despite belonging to the same group of CI users CI6 had a loudness function, for the S_0N_{-60} condition, resembling those of the CI users that were able to rate the loudness of the masked sinusoid. This loudness function has a vertical lower slope, revealing a masked threshold of about 72 dB, and a rather flat upper slope. Besides CI user CI7, the second group of CI users also comprises CI4 and CI5. Within this group, the most obvious differences were found for the S_0 condition. The function estimated for CI4 consists of two lines with about equal slopes and a threshold of 27 dB. For CI user CI5 the loudness function covers a smaller level range, revealing a threshold at 55 dB and an uncomfortable level at about 100 dB. The masking conditions lead to similar shapes of the loudness functions for the three CI users although individual magnitudes differ. For CI users CI5 and CI7 thresholds were in the range from 59 to 70 dB with the lowest threshold (59 or 62 dB) for the S_0N_0 condition. Thresholds of CI user CI4 were between 49 and 60 dB with the lowest threshold found in the S_0N_{60} condition.

6.4 Discussion

Normal-hearing listeners

The average data of the normal-hearing listeners shown in Fig. 6.3 indicate an average spatial release from masking of 3 dB for the S_0N_{60} and of 7 dB for the S_0N_{-60} condition at threshold. The difference in threshold might occur due to individual differences between the listeners as already found in Chapter 5 of this thesis. Some of the listeners show, to a greater or lesser extent, a greater advantage for the S_0N_{-60} condition (e.g. listener NH2), others for the S_0N_{60} condition (listener NH4). However, the SRM of 3 to 7 dB found in this study is in agreement with earlier studies on SRM in normal-hearing listeners. Saberi *et al.* (1991), investigated the detectability of a click train in a broadband masking noise for 36 signal locations as a function of seven masker locations. For a spatial setup comparable to the S_0N_{-60} condition of this study, the authors obtained a release from masking of about 7 to 9 dB compared to the baseline condition. Similar results were found by Hawley *et al.* (2004) who masked spoken sentences with spectrally matched speech-spectrum shaped noise. In a non-speech pattern identification task Kidd *et al.* (1998) found SRMs of about 2 to 5 dB for a signal-masker separation of 60° that were also within the range of the results of this study. Note that other studies which revealed binaural masking level differences of up to 20 dB (e.g., van de Par and Kohlrausch, 1997) conducted the measurements using headphones and compared a $S_\pi N_0$ release condition to a S_0N_0 baseline condition. The phase difference between the conditions in those setups is considerably greater as compared to the present setup, and hence leads to greater level differences. To a lesser extent, the rather low SRM could be explained by the fact that the masker was simultaneously gated with the signal. Bremen and Middlebrooks (2013) measured the SRM for synchronous and asynchronous signal-masker onsets. The results revealed considerably smaller SRMs for the synchronous than for the asynchronous condition.

The categorical loudness scaling procedure used in this study not only allows for the investigation of the SRM at threshold but also at suprathreshold levels. On average, the suprathreshold SRM can be observed up to about 20 to 25 dB above the masked threshold in the S_0N_0 condition. Earlier studies on binaural unmasking (Townsend and Goldstein, 1972) and binaural masking level difference (Soderquist and Schilling, 1990) also investigated masking release effect at suprathreshold levels. In agreement with the results of this study, in both investigations, the effects disappeared at about 20 dB above the masked threshold in the baseline condition.

CI users

Generally, on the basis of the results shown in Figure 6.5, CI users can be divided in two groups. The results of CI users CI1, CI2, CI3, and CI6 (see Fig. 6.5) indicate that one group of CI users was not able to distinguish between the masking noise and the sinusoidal target at low levels. CI user CI2 even reported that he did not hear a distinct difference between the target and the masker at all levels. This is confirmed by the resulting loudness functions: for the three masking conditions loudness ratings larger as or equal to 10 CU (see lower panel in Fig. 6.4). Similar results were found for CI user CI3, whereas he perceived a clear difference between the sinusoid and the noise masker. In spite of that, he reported to hear the tone in the masking noise even at very low levels. For the second group of CI users, at least higher thresholds for the masking conditions were found as for the S_0. However, no spatial release from masking was found for this group of CI users.

Regarding the individual results of the CI group that is not able to distinguish between target and masker at low levels (e.g. lower panel of Figure 6.4) it is debatable whether the fitting function introduced by Brand and Hohmann (2002) is suitable for those data. Due to the definition, the transition region between the upper and the lower fitting function is always located at a loudness of 25 CU. However, concerning the results in the lower panel of Figure 6.4, as well as in Figures 9.16 to 9.13 in the Appendix, the intersection point might be shifted to lower levels. A better fit could probably be achieved by introducing an additional parameter as proposed in Anweiler and Verhey (2006).

When measured using similar stimulus conditions as for common studies in normal-hearing listeners, BMLDs in CI users were about 9 dB when stimulated directly with two pairs of electrodes (Long *et al.*, 2006). Admittedly, considering the results of binaural listening tasks there are performance differences in CI users under those ideal listening conditions as compared to more realistic conditions (e.g. SRM of 3 to 4 dB as found in Loizou *et al.*, 2009). The setup in this study reveals a more realistic listening condition due to maintaining realistic spatial listening cues by means of signal presentation in free-field. This fact might, to a certain extent, account for the rather poor performance of the CI user in the present listening tasks. Additionally, in literature, SRM in normal-hearing listeners was found to be larger for spectrally similar than for spectrally separated signal and masker (Hawley *et al.*, 2004), where as for CI users the exact opposite seems to be true (Loizou *et al.*, 2009). The signals used for this study were a 500-Hz sinusoidal target and an one-octave wide Gaussian noise with a geometric center frequency of 500 Hz. Based on the literature data a smaller SRM is expected for the CI users than for the normal-hearing listeners in agreement with the data of this study.

6.5 Summary

In the present study, spatial release from masking was investigated at threshold and suprathreshold levels in normal-hearing listeners and CI users using a sinusoidal target signal and a narrow-band masking-noise. For the normal-hearing listeners results at threshold (SRMs of 3 to 7 dB) are within the range of earlier studies using comparable measurement setups. The benefit in the masking release condition can also be found at suprathreshold levels up to about 20 to 25 dB above the masked threshold in the baseline condition. This is also in agreement with the results of previous studies that compared the loudness of a sinusoid in a masking release condition to that of a sinusoid in a baseline condition. For the CI users no release from masking due to spatial separation of the signal and the masker was observed. In fact, due to the results, the CI users were divided in two groups, from which one generally had issues in distinguishing between the masking noise and the sinusoidal target at low levels.

7 Free-field measurement of minimum audible angle and movement angle in cochlear implant users

7.1 Introduction

The ability to localize sound sources influences the spatial orientation and helps separating different sound sources, such as a talker at a road with heavy traffic. A common method to determine the precision of auditory localization is to measure the minimum audible angle (MAA, Mills, 1958), i.e., the just noticeable difference in azimuth at which a listener is able to differentiate between two stationary signals. In the literature the localization ability of the auditory system has mostly been treated from this static perspective (e.g., Mills, 1958; Litovsky, 1997; Perrott and Pacheco, 1989; Perrott et al., 1989; Perrott and Saberi, 1990; Strybel and Fujimoto, 2000). However, many natural sound sources are not static, but are moving either due to the movement of the sound source itself or the movement of the body or head of the listener. Movement perception was usually quantified by a minimum audible movement angle (MAMA, e.g., Chandler and Grantham, 1992), i.e., the smallest angular distance a source must traverse in order to be perceived as a moving source. A common assumption is that the knowledge of the localization of stationary sound sources can directly be transferred to the localization performance of moving sound sources (Perrott and Musicant, 1977). The perception of those dynamic sound sources has thus merely been reported a few times between the early seventies and the early nineties. In most of these studies one or two loudspeakers were mounted on a rotating boom to realize a moving sound source (e.g. Perrott and Musicant, 1977; Perrott and Tucker, 1988; Perrott and Marlborough, 1989; Saberi and Perrott, 1990; Chandler and Grantham, 1992) or a loudspeaker was mounted on a cart in front of the listener (Harris and Sergeant, 1971). Thus, in these studies a moving sound source was generated by physically moving a loudspeaker in space. Some studies simulated movement by increasing or decreasing interaural time differences (ITDs) and interaural level differences (ILDs) for successive dichotic stimuli that were presented via earphones (Altman and Viskov, 1977) or by fading in and out adjacent loudspeakers (Grantham, 1986). According to Saberi and Perrott (1990) real motion is preferable compared to simulated motion, but could lead to unintentional external noise. Thus, for them, the simulated motion seemed to be a practical alternative. Perrott and Musicant (1977) used a two-speaker system mounted on a rotating boom with the corresponding motor placed outside on top of the measurement chamber. According to the authors, their system was noiseless up to a velocity of 180°/s. Above that velocity wind noise became noticeable. Perrott and Tucker used a similar setup and reported that the boom-motor-system was inaudible for velocities below 20°/s. For higher velocities the motor noise became evident but never exceeded 25 dB SPL. In addition, it was uncorrelated with the position of the source and thus did not affect the measurements. It is, however, likely that the sound level correlated with the velocity of the sound source which limits the usage of this setup for studying the ability to detect or discriminate moving sound sources. From this perspective, virtual sound sources presented via headphones seem to be the preferable option as they provide a controlled sound field. Nevertheless, the virtually constructed moving sound sources by changing ITDs and ILDs over time could also lead to unrealistic spatial hearing impressions. Unwanted head movements, for instance, do not lead to a change in the hearing impression, as the sound field is build up inside the head, at least if the setup does not account for spontaneous head movements. Furthermore, headphone measurements do often not comprise modifications of the spectra by the upper body, head and pinnae, e.g., head related transfer functions (HRTF). This also leads to unrealistic hearing impressions due to front-to-back confusion. An improvement in auditory space perception using HRTFs was, e.g., shown by Plenge (1974) and Wightman and Kistler (1989). Despite improved HRTF conduction procedures (e.g. Seeber and Fastl, 2003), realistic headphone measurement still take more time than external stimulations via loudspeaker arrays.

Additionally, headphones are inappropriate for studies with listeners being in need of a hearing devices such as cochlear implant (CI) users. CIs are the only accepted intervention that can restore partial hearing to adults and children suffering from severe-to-profound hearing loss through electric stimulation of the auditory nerve. Although great progress has been achieved in improving the speech perception performance of most CI users, much remains to be done in areas beyond speech perception, e.g., in spatial hearing. During the last decade, localization ability of CI users in free-field was tested in several studies (e.g. van Hoesel *et al.*, 2002; van Hoesel and Tyler, 2003; Nopp *et al.*, 2004; Schoen *et al.*, 2005; Senn *et al.*, 2005; Verschuur *et al.*, 2011; Grantham *et al.*, 2007; Seeber and Fastl, 2008). Most of the studies investigated the sound direction identification or localization discrimination in CI users. In these studies listeners were asked to identify from which loudspeaker they considered

the sound to have originated, either by pointing at or calling
out the number of the loudspeaker. The number of loudspeakers raged from seven (Schoen *et al.*, 2005) to 43 (Grantham
et al., 2007) in a range from ±50 to ±50° of arc. In the latter study only 17 loudspeaker between -80° and +80° of arc
were actually playing back the signals. The Duration of the
signals lasted from 200 ms (Grantham *et al.*, 2007) to over 1 s
(Nopp *et al.*, 2004). Localization errors leaded from less than
5° to over 50° depending on several parameters such as stimulus type, presentation angle, and listeners. Only one study
(Senn *et al.*, 2005) tried to investigate the lower limits of spacial discrimination in CI users by measuring the MAA. Instead
of a setup comprised of several loudspeakers the authors used
one loudspeaker on a rotating boom. The same procedure was
already used in Haeusler *et al.* (1983) for normal-hearing and
hearing impaired listeners. With their setup they determined
MAAs in the frontal hemisphere of 4 to 8° which were comparable to the results of normal-hearing listeners (1 to 4°) as well
as to the results of van Hoesel and Tyler (2003).

Regarding the previous, the most realistic attempt that is suitable for all kinds of listening groups seems to be an external stimulation with a controlled sound field. In this study a
method is presented which attempts to comprise the advantages
of realistic external sound sources and a controlled sound field
without the usage of physically moving sources or manipulating
the ITDs and ILDs of the presented sounds. Instead, 31 loudspeakers were arranged in a semicircle and signals were generated by superimposing the signals of consecutive loudspeakers.
The applicability of this setup for measuring the minimum audible angle and minimum audible movement angle in different
listening groups was tested with normal-hearing listeners and
CI users.

7.2 Materials and methods

7.2.1 Setup

Figure 7.1 shows a schematic representation of the setup. Thirty-one equidistant self-powered loudspeakers (Genelec 6010A) were mounted on a semicircle with a diameter of 3.5 m at a height of 1.2 m. The loudspeakers were positioned from -90° to +90° in 6°-steps including a center loudspeaker at 0° of arc. The setup was placed in a sound-reduced booth. All stimuli were generated digitally using MATLAB at a sampling rate of 44.1 kHz. To play back signals via more than two channels the Playrec utility was used which provides versatile access to sound cards using PortAudio. The stimuli were converted from digital into analog signals using four 8-channel RME ADI-8 QS and presented via the 31 loudspeakers. The listeners were seated in the center of the semicircle with their ears or, in the case of the CI users, their processor microphones positioned at the same height as the loudspeakers.

7.2.2 Stimulus generation

In a first step the time function of the angle is defined. In the present study this was a stationary or a moving source. The following equation gives the angle of the maximum of the sound source as a function of time t for a movement of duration of T with a constant circular velocity v of the sound source:

$$\varphi(t) = \varphi_b + v \cdot t \text{ with } v = \frac{\varphi_e - \varphi_b}{T} \text{ and } 0 \leq t \leq T \quad (7.1)$$

The angle is φ_b is the launching angle and φ_e is the angle at the end of the movement. By setting $\varphi_e = \varphi_b$ this equation also describes a stationary sound source. Note that, $\varphi(t)$ does not necessarily have to be determined by a movement with a constant angular velocity v but can also assume each type of accelerated movement.

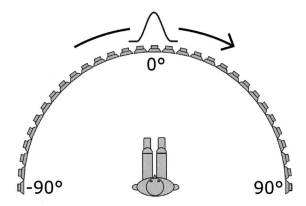

Figure 7.1: General setup for measurement of static and dynamic angle discrimination: 31 equidistant self-powered loudspeakers mounted on a semicircle from -90° to +90° in 6°-steps including a center loudspeaker at 0° of arc. The blues curve reveals the shape of the moving source.

In general, the sound source was generated by superimposing the signals of consecutive loudspeakers. Thus at each point in time t the gains of a number of loudspeakers depending on the width of the sound source were set. In the current experiments the shape of the sound source was given by the following raised-cosine window:

$$a(t, n) = \left(0.5 \cdot \left[1 + cos \left(\frac{2 \cdot \pi}{2 \cdot w_{src}} \cdot \textbf{min} \left(w_{src}, |\varphi(t) - \varphi_{ch}(n)| \right) \right) \right] \right)^{0.5}$$

$$(7.2)$$

where w_{src} is half the width of the sound source, $\varphi_{ch}(n)$ is the angle of the nth loudspeaker and $\varphi(t)$ the angle of the maximum of the sound source for a certain point in time. For the movement used here it is determined by equation 7.1. The ab-

174

solute value of the minimum between w_{src} and the difference between $\varphi(t)$ and $\varphi_{ch}(n)$ ensures that the amplitudes of those loudspeakers that are not within the range of the source are set to zero. Thus, the number of running loudspeakers and their corresponding amplitude at a certain point in time is determined by the source width $2 \cdot w_{src}$. The square root of the term is used to ensure that intensities of the different loudspeakers are added. Equation 7.2 leads to a matrix where each column describes the time course of the amplitude of a specific loudspeaker.

As an example, Figure 7.2 shows how a movement with a constant angle velocity of the sound source is realized. The right panels show the matrix used for the array for a sound source moving from -90° to +90° in 4 s. The left panels show the amplitudes for the loudspeaker for two different points in time t (black) and $t + \Delta t$ (grey). The upper panel shows the movement for a 60° wide source, the lower panel shows that for a 20° wide source as used in the current psychoacoustical experiment. For the 60° wide source, nine loudspeakers are playing back the signal at the two instances in time shown in the left panel, each with a different amplitude. For the 20° wide source shown in the lower panel at the same points in time only four loudspeakers are playing the sound. For both points in time, the amplitude maximum of the source lies in between two loudspeakers.

Figure 7.3 shows the modulation spectrum of an one-octave wide noise with a constant angular velocity v of $45°/s$ and a duration T of 4 s. The different lines indicate different source widths $2 \cdot w_{src}$: 2, 4, 6, 12, 20 and 60°. The brightness of grey increases with increasing source width. For source widths smaller than 12°, there is a distinct peak in the modulation spectrum at a frequency of about 7.5 Hz. There are also peaks for multiples of this frequency but they are less pronounced. The peaks decrease in height with increasing source width. This decrease is more prominent for multiples 7.5 Hz. The unintended modulation originates from the limited number of loudspeakers in

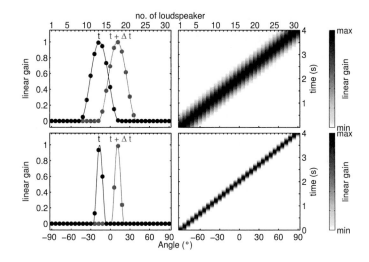

Figure 7.2: Realization of a linear movement of the sound source within the loudspeaker array for source widths of 60° (upper panels) and 20° (lower panels), respectively. **Left:** Shape and position of the sound at time t (black) and $t + \Delta t$ (grey), the dots represent the different loudspeakers with the given gain for the two different points in time. **Right:** Time course of the moving sound source.

the array. For source widths smaller than 12°, the instantaneous level depends on the position of the source relative to the loudspeakers. Only if the position of the center of the source equals the exact position of a loudspeaker the source has the intended level. For all other source positions the amplitude is lower than intended. This results in a periodic level fluctuation in the case of a constant angular velocity. The frequency is equal to the ratio of constant angle velocity and the distance of

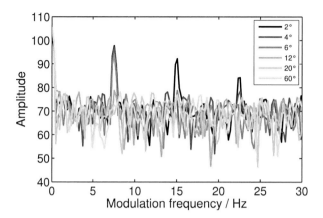

Figure 7.3: Modulation spectrum of a signal with a source width of either 2, 4, 6, 12, 20 or 60° using colors with increasing brightness from black (2°) to light grey (60°). The distinct peaks indicate unwanted periodic modulations. These are observed for source widths smaller than 12°.

the loudspeakers. For source widths equal to or above 12° this level fluctuations no longer occur since at any point in time several loudspeakers play back the signal ensuring the the intensity does not depend on the source position.

For the psychoacoustical experiments of the present study the sound was an one-octave wide Gaussian noise with a geometric center frequency of either 500 or 2500 Hz. The noise was gated with 10-ms raised-cosine windows at on and offset. The level of the noise was set to 65 dB SPL for normal-hearing listeners and to a most comfortable level for CIs. This level was determined individual for each CI user.

7.2.3 Calibration

Equalization and calibration were done in two steps: In the first step all loudspeakers were equalized in amplitude by adjusting the frequency responses of the single loudspeakers to an average frequency response. Adjusting the loudspeakers to an average loudspeaker ensures that the required amplification for the equalization at each frequency is kept within reasonable bounds. For the equalization, a broad-band Gaussian noise was played back from each loudspeaker separately and recorded with a prepolarized 1/2-inch free-field microphone unit (Bruel & Kjær 4188-A-021, consisting of a microphone type 4188 and a pre-amplifier type 2641) and a subsequent amplifier (Bruel & Kjær 1704-A-002). The played back and recorded signals were then used to calculate the frequency responses of the corresponding loudspeakers. To this end, the signals were cross correlated and the maximum of the cross correlation was used to correct the time delay between the signals. Subsequently, the signals were transferred into the frequency domain and, for each loudspeaker separately, the recorded signal was divided by the played back signal, resulting in the transfer functions for each loudspeaker. Those transfer functions were then transformed back into the time domain using a Fast Fourier Transform and windowed in order to restrict the impulse response to that for the direct sound. Subsequently, the windowed impulse responses were transformed into the frequency domain and the average frequency response was calculated. The equalization function was calculated for each loudspeaker separately by dividing the average frequency response by the respective loudspeaker frequency response. The correction was only performed in the range from 150 to 10000 Hz. Outside this range the correction values were set to one. All correction functions were then stored in one correction matrix for the whole loudspeaker system.

In the second step, the signal or the psychoacoustical experiment was calibrated. For this purpose, at first the microphone unit was calibrated using a calibrator (Bruel & Kjær 4231) that emitted a sinusoid of 94 dB SPL at 1000 Hz. Subsequently, the signal was generated as described in the previous section. The source width was 20°. To equalize the loudspeakers, the measurement signal was transferred into the frequency domain an multiplied by the correction matrix and transferred back into the time domain. The measurement signal was then played back from the frontal position (0°) and recorded with the microphone. The internal representations of the calibration signal cal_{int} and of the recorded measurement signal rec_{int} were then used to determine the calibration level which was used to set the correct level for the measurement process.

7.2.4 Procedure

Minimum audible angles and minimum audible movements angles were measured using similar measurement procedures: The angle resolution for the CI users was determined using a constant stimulus alternative forced-choice (AFC) procedure with 27 repetitions, including two test repetitions, for each value of experimental parameter. The signal duration was set to 1000 ms. For the normal-hearing listeners an AFC procedure with adaptive signal-level adjustment (1-up 2-down) leading to the 70.7 percent-correct value on the psychometric function was used. In contrast to the signals used for the CI users, duration was set to 500 ms to ensure appropriate overall durations. To avoid jumps in angle (from positive to negative angles or the other way round) for signal presentations close to the center (0°), especially in the MAA task, the total angle φ_{tot} was expressed in dB using the following equation:

$$\varphi_{tot}(dB) = 20 \cdot log10 \left(\frac{\varphi_{tot}\left(°\right)}{180°} \right) \qquad (7.3)$$

The initial step size was set to 4 dB and reduced by 1 dB after each upper reversal to a final step size of 2 dB. The mean of further six reversals with this step size was taken as estimate for one run. The final estimate of the movement detection threshold was taken as the mean over three such runs. The normal-hearing control group also performed the measurements using a constant stimulus procedure for two values of experimental parameter.

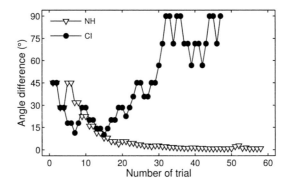

Figure 7.4: Typical adaptive track for a CI user (filled black circles) and a normal-hearing listener (open triangles).

Different procedures for the two groups of listeners were used due to difficulties with the adaptive procedure when measured with the CI users. Exemplary, Figure 7.4 depict the results of one adaptive track for one CI user and one normal-hearing listener. The results show, that at the beginning of an adaptive track, CI users are able to perform the adaptive task. However, during the course of the track they apparently loose their listening cue. The adaptive procedure only allows for attaining the limits of the measurement parameters three times. Since the upper limit is restricted to 90° due to the dimensions of the semicircle, the track for the CI user is skipped when an angle of 90° is reached for the fourth time.

Due to the change in the measurement procedure, in CI users performance in static and dynamic angle discrimination is measured rather than *minimum* audible (movement) angles. However, the terms MAA and MAMA are retained unchanged during this chapter to avoid any possibility of confusion.

Minimum audible angle

The minimum audible angle was measured in the frontal hemisphere using a 3-AFC procedure. Within this 3-AFC task the experimental parameter was the difference in angle between the reference and the test signal. Signals were set symmetrically around the center loudspeaker with the reference signals shifted to the right and the test signal to the left in the same distance. Listeners had to indicate which of the three signals was further to the right side by pressing the according key on the keyboard. For the constant stimulus procedure performed with the CI users the angles passed by the moving signal were set to 90, 45, and 20° of arc. The data for the normal-hearing control group were determined using the adaptive procedure and the constant stimulus procedure for angles differences between test and reference signals of 20 and 45°.

Minimum audible movement angle

The minimum audible movement angle was also measured in the frontal hemisphere using a 2-AFC procedure. Within this 2-AFC task the experimental parameter was the angle passed by the moving signal. The reference signal was set to the center loudspeaker at 0° of arc and the test signal moved symmetrically around this frontal position with velocities of either 90 or 30°/s. The length of the signal was defined by the velocity and the angle passed by the moving signal. Listeners had to indicate which of the two signals had moved by pressing the according key on the keyboard. For the constant stimulus procedure

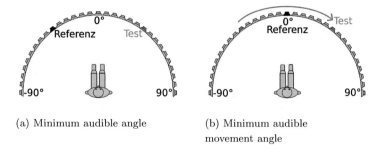

(a) Minimum audible angle

(b) Minimum audible movement angle

Figure 7.5: Measurement setup for the measurement of MAA (left) and MAMA (right). **MAA**: Reference signal was presented from the left hemisphere, test signal from the right. **MAMA**: Reference signal was presented from the front, test signal either moved from left to right or right to left.

performed with the CI users the angles passed by the moving signal were set to 90, 45, and 20° of arc. The data for the normal-hearing control group were determined using the adaptive procedure and the constant stimulus procedure for angles differences between test and reference signals of 20 and 45°.

7.2.5 Listeners

Five bilaterally implanted CI users participated in the study. The CI users CI 1, CI 3, CI 4, and CI 5 used MED-El® CIs (models C40+, Pulsar, Sonata, Concerto) and were fitted with OPUS 2 sound processors. CI 2 (45 years, male) was implanted on both sides contemporary with Nucleus® CIs (model CI24RE) and provided with CP910 sound processors. Detailed listener demographics are shown in Table 7.1.

Table 7.1: CI user demographics

CI user	Sex	Ear	Age at implant	Device	Processor	Strategy	Duration of use (years)
CI 1	f	L	34 (40)	C40+	OPUS 2	FSP	19
		R	43	Pulsar	OPUS 2	FSP	9.5
CI 2	m	L	44	CI24RE	CP910	ACE	1
		R	44	CI24RE	CP910	ACE	1
CI 3	m	R	49	C40+	OPUS 2	FSP	12
		L	52	Pulsar	OPUS 2	FSP	8.5
CI 4	m	R	47	Concerto	OPUS 2	FS4	3.5
		L	48	Concerto	OPUS 2	FS4	2.5
CI 5	m	L	50	Sonata	OPUS 2	FS4	3
		R	49	Sonata	OPUS 2	FS4	<1

Eleven normal-hearing adults (3 males, 8 females) at ages between 21 and 34 years were enrolled in the study. All had hearing thresholds ≤15 dB HL at standard audiometric frequencies between 125 and 8000 Hz.

7.3 Results

7.3.1 Minimum audible angle

Figure 7.6 shows the results for the measurement of the minimum audible angle. The individual results of the CI users (circles) are shown in the upper four panels and the bottom left panel. The corresponding average data (circles) and the mean data of the normal-hearing control (triangles) group are shown in the lower right panel. Open symbols indicate the results for a center frequency of 500 Hz, filled black symbols for a center frequency of 2500 Hz.

All listeners except of two CI users completed the MAA task. The respective CI users (CI 2 and CI 5) could not perform the condition with a center frequency of 500 Hz. For a center frequency of 2500 Hz (filled circles) the results for CIs 1, 3 and 5 reveal a similar pattern of percentage correct values, although the individual magnitudes differed. In general, values decreased with increasing angle difference with a steeper slope towards smaller angle differences. A comparable, although more pronounced pattern can also be found in the results of the normal-hearing listeners (lower right panel). For CI 2 and CI 4 percentage correct values also decreased with decreasing angle difference. However, for CI 4 the slope remained constant in the whole measurement range, whereas for CI 2 the slope varied between the different measurement points with comparable results around 50% correct for angle differences of 5 and 10°. For a center frequency of 500 Hz (open circles) results were only available for CI 1, CI 3, and CI 4. CI 3 showed similar results as for 2500 Hz, starting with the same value of about 95% correct

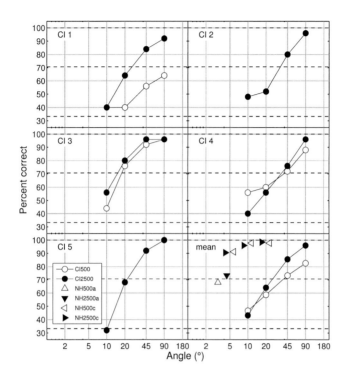

Figure 7.6: Individual and average results of CI users (circles) and average results of the normal-hearing listeners (lower right panel, triangles). Open symbols indicate the results for a center frequency of 500 Hz, filled black symbols for a center frequency of 2500 Hz. Upward and downward pointing triangles mark the results determined with the adaptive procedure (NH500a and NH2500a), left and right pointing triangles for the constant stimulus procedure (NH500c and NH2500c).

at 90° and steeper slopes towards smaller angle differences. For CI 1 percentage correct values were significantly lower for 500 compared to 2500 Hz with values not exceeding 65% and values close to chance level at angle differences of 10 and 20°. For CI 4 percentage correct values for 500 Hz compared to 2500 Hz were lower at 90°, about the same at 45 and 20° and better at 10°, although still following the general trend of decreasing performance with decreasing angle difference. The general trend is also reflected in the average results. Performance is better for the higher center frequency for angle differences of 45 and 90°. For 20 and 20° performance is about the same for both types of stimuli.

The upward and downward pointing triangles mark the MAAs measured with the adaptive procedure in the normal-hearing listeners. For a center frequency of 500 Hz (open triangles) a MAA of 3.2° was found whereas for a center frequency of 2500 Hz (filled triangles) the MAA of 4.4° was only about 1° higher. The right and leftward pointing triangles denote the results for the constant stimulus measurements. At a fixed test angle of 20° the results amounted to 97.8% correct for a center frequency of 500 Hz and to 98.5% correct for 2500 Hz. The percent correct values for a test angle of 10° stayed about the same compared to the wider test angle for both center frequencies: 97.8% at 500 Hz and 96.0% at 2500 Hz. For the lowest test angle of 5° percent correct values decreased for both center frequencies by 6.5 and 5.5°, respectively. Thus, for the normal-hearing listeners results were in the same range for both center frequencies with a slightly steeper loudness function for the higher frequency.

7.3.2 Minimum audible movement angle

Figure 7.7 shows the individual results for the MAMA task. The results of the CI users (circles) are shown in the upper four panels and the bottom left panel. The corresponding average

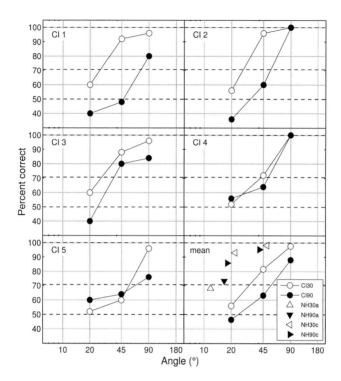

Figure 7.7: Individual and average results of CI users (circles) and average results of the normal-hearing listeners (lower right panel, triangles). Open symbols indicate the results for a velocity of $30°/s$, filled black symbols for a velocity of $90°/s$. Upward and downward pointing triangles mark the results determined with the adaptive procedure (NH30a and NH90a), left and right pointing triangles for the constant stimulus procedure (NH30c and NH90c).

data (circles) and the mean data of the normal-hearing control (triangles) group are shown in the lower right panel. Open symbols indicate the results for a velocity of $30°/s$, filled black symbols for a velocity of $90°/s$.

CI users 1, 2 and 3 show a similar pattern for the percent correct values as a function of test angle and velocity. In general, the percentage correct was largest for wide angles passed by the signal and the lower velocity and lowest for narrow angles and the higher velocity. CI 1 and CI 3 showed a better performance for a velocity of $30°/s$ over all angles. The same applies for CI 2, despite for a movement angle of $90°$ at which the percentage correct was 100% for both velocities. For CI 3 the results for movement angles of 45 and $90°$ were markedly above chance level, whereas the results for a movement angle of $20°$ (40 and 60%, respectively) were close to the chance level of 50%. For CI 1 and CI 2 the results for both velocities and a movement over $90°$ and the result for $45°$ with an velocity of $30°/s$ amount to 80% or higher whereas the other results were around chance level. For CI 4 and CI 5 the resulting measurement curves had a different pattern. CI 4 reached 100% correct for a movement angle of $90°$ at both velocities. For a movement angle of $45°$ values decreased to about 72 and 64% correct for the lower and the higher velocity, respectively. For a movement angle of $20°$ values were close to chance level for both velocities. For the widest angle passed by the moving signal, CI 5 reached a percent correct value close to 100% for the lower velocity, whereas for a velocity of $90°/s$ the percent correct value was approximately 75. At movement angles lower than $90°$ CI 5 performed better for the higher compared to the lower velocity. For a movement angle of $20°$ performance was close to chance level for a velocity of $30°/s$ and at about 60% correct for a velocity of $90°/s$. The differences between the two center frequencies tested were most distinct for CI 1, followed by CI 2, 3 and 5. Hardly any differences were found for CI 4. However, the average results reveal a pattern similar to results of CI 1 to CI 3: for a movement angle of 90° and a velocity of 30°/s the percent correct value was close to 100. With increasing velocity and decreasing movement angle the percent correct values decreased until they reach chance level for a movement angle of 20° for both velocities.

The upward and downward pointing triangles mark the MAMAs measured with the adaptive procedure in the normal-hearing listeners. For a velocity of 30°/s, a MAMA of 11.8° was found whereas the MAMA was about 5° higher (16.7°) for a velocity of 90°/s. The right and leftward pointing triangles denote the results for the constant stimulus measurements. For a movement angle of 45° 98.2% of the responses were correct for a velocity of 30°/s and 95.3% for 90°/s. The percent correct values for a movement angle of 20° were lower compared to those for the wider movement angle for both velocities. For 30°/s it decreased by 5° and for 90°/s by about 10°.

7.4 Discussion

7.4.1 Applicability of setup and signal generation

Smoothly moving signals can be achieved, when an appropriate source width is chosen. This is determined by the distance between consecutive loudspeakers. Since the stimuli were generated by superimposing the signals of the loudspeakers, the minimum allowed source width is twice the distance between two consecutive loudspeakers. A source widths smaller than twice the loudspeaker distance leads to a wobbling hearing impression due to fluctuation in level, as indicated by distinct peaks in the modulation spectrum (Fig. 7.3).

The general possibility to use this setup and signal generation for the measurements of the minimum audible angle and movement angle was tested in eleven normal-hearing listeners. For the MAA results were slightly higher, than the results for normal-hearing listeners obtained in earlier studies with different types of stimuli (1°, e.g., Mills, 1958; Perrott et al., 1989; Litovsky, 1997). However, Mills (1958) reported higher MAAs for stimuli with frequencies around 2500 Hz (about 3°) as compared to stimuli with frequency contents below 1000 Hz (about 1°). This tendency can also be observed in the present data with MAAs of 3.2° at a center frequency of 500 Hz and 4.4° at a center frequency of 2500 Hz. Other studies reported MAAs ranging from about 1 to 4 or 5° for their individual listeners (Haeusler et al., 1983; Grantham et al., 2003) that were within the same range as the average MAAs found in this study. Additionally, Chandler and Grantham (1992) who considered the MAAs to be a special case (velocity of 0°/s) of the MAMAs, estimated MAAs from their MAMA results. This estimation lead to average MAAs of 2.6° for a frequency of 500 Hz and 8.4° for 3000 Hz.

In agreement with earlier studies on movement perception which were performed using rotating booms or headphones and manipulated ITDs and ILDs, the present results for the MAMA in normal-hearing listeners show an increasing detection threshold for movement as the source velocity is increased. Compared to the results of Chandler and Grantham (1992) or Altman and Viskov (1977), the thresholds were in the same range. Chandler and Grantham (1992) determined thresholds of about 15° for a velocity of 45°/s and of about 10° for a velocity of 90°/s. Other studies (e.g. Perrott and Musicant, 1977; Perrott and Tucker, 1988; Perrott and Marlborough, 1989; Saberi and Perrott, 1990), however, consistently revealed lower thresholds at about these velocities.

The differences between the MAAs and MAMAs determined in the present study and those obtained in earlier studies might be due to other differences in the measurement parameters such as frequency range, duration and temporal structure or even presentation level rather than the setup itself. Hartmann and Rakerd (1989) also reported differences in the MAA due to different measurement paradigms which lead to different decision strategies of the listeners and thus to different values for the MAA.

7.4.2 Minimum audible (movement) angles in CI users

In general, the results for the minimum audible angle task were poorer in CI users as compared to the normal-hearing listeners. The tendency of poorer performance with decreasing angle difference between the test and the reference signal, however, can be observed in both groups of listeners. Two of the CI users (CI2 and CI5) were not able to complete the MAA task for the signal center frequency of 500 Hz as they could not associate the signal with a specific direction in space. They reported a perception of a rather diffuse sound field which made it impossible for them to recognize the direction from which the sound originated. One major point they have in common is the duration of use in at least one ear. CI2 was implanted simultaneously on both side less than a year before the measurements. The same applies for the right ear of CI5. Thus one reason for the difficulties at the lower center frequency might be the missing practice. However, the results for a center frequency of 2500 Hz were not noticeably worse as compared to the other CI users. At least for CI5, the performance can be explained by a monaural auditory cue referred to as the head shadow effect (HSE, van Wanrooij and van Opstal, 2004). Particularly for higher frequencies (wavelength rather small compared to the dimensions of the head), the sound is attenuated by the head as a function of direction. Depending on frequency, this attenuation can lead to level differences up to 25 dB (Nava *et al.*, 2009). For low frequencies however, the wavelengths are rather large compared to the dimensions of the head and thus do not produce significant level differences in most cases (Verschuur *et al.*, 2011). In fact, this could also partially explain the results of CI2 and CI1, as the monaural cue HSE also holds true for binaural hearing leading to interaural level differences (ILDs). For lower frequencies there is a greater dependency of localization ability on interaural time differences (ITDs) in phase or signal onset. Due to the the signal processing in the cochlear implants (e.g., envelope extraction by means of low-pass filter, Wilson and Dorman, 2008), temporal information is only coarsely coded. Despite CI2, all CI users were fitted with FSP or FS4 strategies which preserve the temporal fine structure of the signals at least in the lower one to four channels (Hochmair *et al.*, 2006). This could also explain the similar performances for the two different stimuli of CI3 and CI4. Despite higher MAAs for the present group of CI users, the findings that the minimum audible angles mainly depend on level cues and less on temporal cues are in line with Senn *et al.* (2005).

Due to better performances with the 2500-Hz stimulus, the 500-Hz stimulus was not included in the MAMA task. Overall, movement detection was poorer in the CI users group than for the normal-hearing listeners. However, the general tendencies (decreasing performance with decreasing angle and increasing velocity) were similar for both groups of listeners. In fact, for the lower velocity and the wider angles, percentage correct values of the CI users were within the range of the normal-hearing listeners whereas for the higher velocity performance drops more rapidly. This is especially true for CI users CI1 and CI2. A possible reason for the rapid drops in performance especially at the higher velocity is the signal duration. As the velocity is kept constant, signal duration decreases with decreasing movement angle from a duration of 1 s for a movement angle of 90° to about 220 ms for 20°. However, Grantham *et al.* (2007) used 200-ms long signals in their source direction identification task and did not report difficulties among the CI users in perceiving signals of this duration.

7.5 Summary

An experimental setup with 31 loudspeakers and a corresponding signal generation algorithm was presented that allows to play back moving signals. Essentially, there are no restrictions regarding the velocity contour of the moving sound source. Since the movement is virtually generated, the realization of accelerated movements as well as changing velocity contours within one stimulus is possible. Additionally, the setup allows for presentation of multiple sound sources simultaneously, e.g., for measurements in more realistic complex sound environments with different sound sources from different directions.

The possibility to use this setup and signal generation for the measurements of the minimum audible angle and movement angle was tested in CI users and normal-hearing listeners. The results of the normal-hearing listeners revealed similarities and differences compared to earlier studies. Due to the lack of respective literature data on MAA and MAMA per se, the results obtained in CI users could hardly be compared to results of localization tasks in earlier studies. However, general trends observed in normal-hearing listeners were also observed in CI users. Despite some differences, the presented measurement setup and the stimulus generation seems applicable for investigations on spatial hearing in normal-hearing listeners as well as in CI users.

8 Summary and Outlook

This thesis investigats the influence of spectral and temporal aspects of loudness perception in normal-hearing listeners, as well as the binaural and spatial listening abilities of normal-hearing listeners and CI users.

The first study of this thesis, Chapter 2, focus on spectral loudness summation by means of pulse trains. In agreement with previous studies (Fruhmann *et al.*, 2003; Verhey and Uhlemann, 2008) it is shown that loudness is influenced by the repetition rate and the duration of the pulses within the sequence. Besides the simultaneous spectral loudness summation, i.e. increasing loudness with increasing bandwidth at a constant intensity, loudness summation was also observed when different frequency components were presented subsequently. This effect can be explained by the assumption, that the specific loudness in one auditory filter builds up rapidly but subsides only slowly (Zwicker, 1969). However, the dependency on the repetition rate and the non-simultaneous spectral loudness summation appear to be opposing effects partially compensating each other. Further, the stimuli used for this study are applied to a loudness model that predicts the duration dependency of spectral loudness summation (eDLM Rennies *et al.*, 2009). While the predictions for short and long pulses (i.e., long and short IPIs) were in line with the measured data, the rather inaccurate predicted transition region might be of interest for further investigations on the eDLM.

Chapter 3 deals with loudness perception on the basis of spectral and temporal variations within one signal. Previous studies showed that the first segment of a stimulus temporally receives the highest weight for loudness judgment (primacy effect, e.g. Oberfeld, 2008), with a higher effect for broadband than for narrowband stimuli (Rennies and Verhey, 2009). This primacy effect was also obtained as can be seen by the results of Chapter 3, whereas a dependency on bandwidth could not be confirmed. Higher weightings on lower frequency bands, as found in the present study, are only in line with one previous study on this topic (Jesteadt *et al.*, 2011). However, deviations to other studies might be explained by the usage of equal loudness instead of equal level for the different frequency components. Besides considering temporal and spectral weightings separately, as a new approach, the influence of spectro-temporal weights, i.e., temporal level variations that are independent for different spectral regions, on loudness perception was tested. The results show that spectral or temporal weights that were estimated in the spectro-temporal weights task do not differ from the weights estimated in the simple spectral or temporal-weights tasks. These findings reveal, that a prediction of spectro-temporal weights on the basis of spectral and temporal weights is possible at high precision. As a continuation of these experiments, an interesting issue would be the application of the stimuli to current loudness models such as the already mentioned eDLM.

The Chapters 4 to 6 focus on the feasibility of the categorical loudness scaling (CLS) procedure for areas beyond clinical application. For one thing, the common measures of loudness are sone and phon, which are also used in loudness models. On the other hand, the loudness functions resulting from CLS experiments are given in categorical units (CU) over presentation level. Thus, a direct comparison between the measured loudness functions in CU and predicted loudness values in sone or standardized noise emission limits in phon is hardly possible. In Chapter 4 indication is given for the relations between the different loudness measures by introducing an equation making possible the transformation of the loudness in sone to loudness in CU. However, the conversion in the opposite direction based on the proposed equation is only achievable using numerical solutions. Further investigations on the relations would thus be an interesting issue for future studies. Furthermore, suprathreshold perception is often investigated by means of time consuming adaptive matching procedures. In the Chapters 5 and 6 the CLS procedure is used to investigate suprathreshold effects of masking. In Chapter 5 the masking ability of an unmodulated noise masker (baseline condition) is compared to that of a temporally modulated masker (masking release condition) using both, a CLS and a matching procedure. The masking release due to spacial separation of target and masking signal is investigated by means of the CLS procedure in Chapter 6. In agreement with the findings of previous studies (e.g., Townsend and Goldstein, 1972; Verhey and Heise, 2012), for both investigations, masking release could be observed for levels up to at least 25 dB above the masked threshold in the baseline condition. In fact, the CLS procedure in Chapter 5 obtaines effects up to 35 dB which is in line with findings of Zwicker and Henning (1991). Masking-release effects at threshold are also in line with studies of similar tasks (e.g., the free-field release from masking in Saberi *et al.*, 1991). Additionally, the study in Chapter 5 confirms the hypothesis by Verhey and Heise (2012) that the partial loudness of a sinusoid in a modulated masker equals that of a sinusoid in an unmodulated masker that was reduced in level by the masking release at threshold. Further insight on those suprathreshold effects might be gained by investigations of neural correlates in EEG, MEG or fMRI studies.

Besides normal-hearing listeners, Chapter 6 investigates partial loudness in CI users. In contrast to earlier studies on masked signals using direct stimulation via research processors, spatial release from masking was tested in a free-field condition. Direct stimulation might provide a more reliable sound field, but compared to free-field measurements it does not include acoustic cues that are available in realistic listening conditions. Generally, the performance in rating the partial loudness of a sinusoid was poorer in CI users as compared to normal-hearing listeners. One group of CI users was not able to distinguish between the masking noise and the sinusoidal target at least at low levels. As the masking noise was spectrally centered around the the sinusoidal target, a stimulation with equal electrodes might account for this effect. However, spacial release from masking was also not found in those CI users that were able to distinguish both signals even at low levels. To gain further insight on the perception of partial loudness in CI users measurements should be extended to multiple signal-masker conditions regarding spectral content and spacial location and separation. So far, the results indicate that much remains to be done in the area of binaural hearing with cochlear implants. Nonetheless, the latest study (Chapter 7) of the present thesis could show good performance in spacial orientation for bilateral CI users.

In Chapter 7 a free-field measurement setup in combination with virtual signal generation is presented. This setup allows for investigations on the perception of static and dynamic sound sources in normal-hearing listeners as well as in CI users. In normal-hearing listeners minimum audible angles and movement angles are obtained that are comparable to those of previous studies using rotating booms or signal presentation via headphones. Comparable measurements revealed poorer performance in CI users. However, general dependencies on frequency content, angle differences, movement angles and angular velocities are similar to that of normal-hearing listeners. For example, performances at lower frequencies were poorer as compared to higher frequencies in most CU users. Due to the lack of respective literature data on MAA and MAMA per se, the results obtained in CI users could not directly be compared to results of localization tasks in earlier studies. In agreement with previous studies (Verschuur *et al.*, 2011; Seeber and Fastl, 2008), the frequency dependency shows that CI users are nonetheless more reliant on interaural level cues and have less access to interaural time cues. For further comparison to literature data, the presented measurements could be expanded, e.g., to lateral hemispheres. Additionally, the perception of accelerated movements could be an interesting issue of future studies, as the presented setup not only allows for movements with constant angular velocities.

9 Appendix

9.1 Average and individual results of the experiment in Chapter 5 for the whole level range

Figures 9.1 and 9.2 show the average and individual loudness functions and the thresholds of the experiments in Chapter 5. In Figure 9.2 each panel shows data of one subject. The data representation is the same in both figures. Different line styles and stars indicate loudness functions or thresholds for the different masking conditions. The level range shown is the whole level range covered in the measurement of this study.

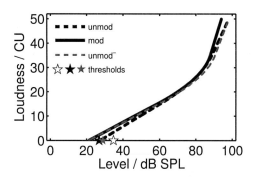

Figure 9.1: Average loudness growth functions for the eighteen subjects participating in the experiments. Different line
styles indicate different masking conditions: mod (solid
line), unmod (black dashed line), and unmod⁻ (grey
dashed line). The black stars indicate the thresholds
for the mod (filled star) and the unmod (open star)
conditions, the grey star indicates the unmod⁻ condition.

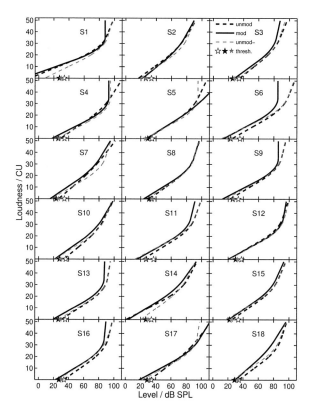

Figure 9.2: Individual loudness growth functions for the eighteen
subjects participating in the experiments. Each panel
shows data of one subject. Different line styles indicate
different masking conditions: mod (solid line), unmod
(black dashed line), and unmod⁻ (grey dashed line).
The black stars indicate the thresholds for the mod
(filled star) and the unmod (open star) conditions, the
grey star indicates the unmod⁻ condition.

9.2 Individual results of the Experiments in Chapter 6

Figures 9.10 to 9.13 show the individual results of the normal-hearing listener (Figs. 9.10 to 9.7) and the CI users (Figs. 9.15 to 9.13) that were not shown in Chapter 6. Each panel denotes data of one measurement condition. Different markers indicate the answers given by the subject in the three different runs. The corresponding fit functions are indicated using different line styles. In each panel, the median loudness function for the respective condition is shown by the solid line.

Normal-hearing listeners

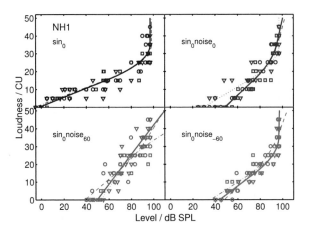

Figure 9.3: Individual results of normal-hearing listener jl in the Experiment in Chapter 6.

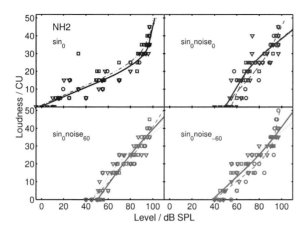

Figure 9.4: Individual results of normal-hearing listener as in the Experiment in Chapter 6.

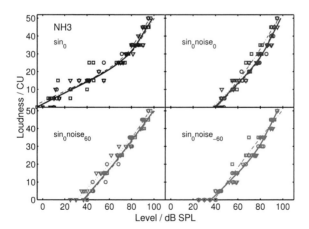

Figure 9.5: Individual results of normal-hearing listener rg in the Experiment in Chapter 6.

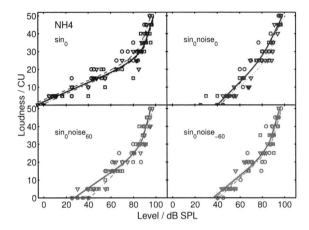

Figure 9.6: Individual results of normal-hearing listener js in the Experiment in Chapter 6.

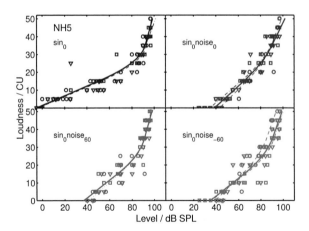

Figure 9.7: Individual results of normal-hearing listener sk in the Experiment in Chapter 6.

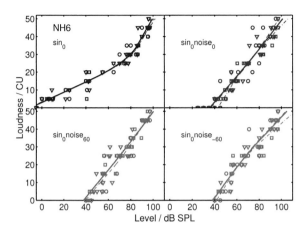

Figure 9.8: Individual results of normal-hearing listener jg in the Experiment in Chapter 6.

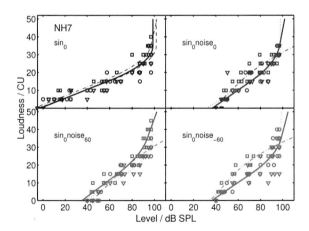

Figure 9.9: Individual results of normal-hearing listener nn in the Experiment in Chapter 6.

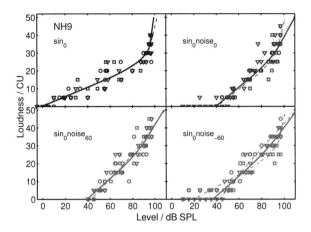

Figure 9.10: Individual results of normal-hearing listener ak in the Experiment in Chapter 6.

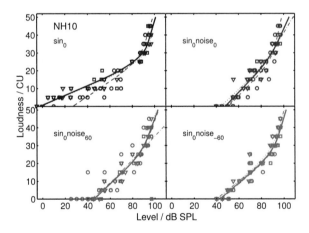

Figure 9.11: Individual results of normal-hearing listener jj in the Experiment in Chapter 6.

Cochlear implant users

On the basis of the results, CI users are divided into two groups. CI users of the first group (Figs. 9.15 and 9.14) were able to rate the loudness of the sinusoid in the noise masker, i.e., when the level of the sinusoid fell belows threshold they rated the sinusoidal target to be "inaudible". The second group of CI users had difficulties to rate or even hear the sinusoid in the masking noise, i.e., even for very low target levels loudness rating never lower that 10 CU (see Figs. 9.12, 9.13, and 9.16).

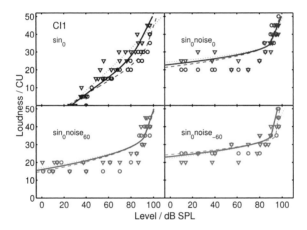

Figure 9.12: Individual results of CI user mr in the Experiment in Chapter 6.

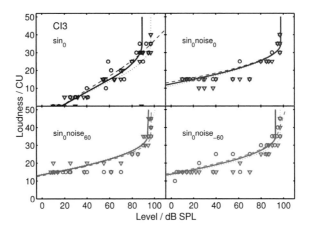

Figure 9.13: Individual results of CI user sr in the Experiment in Chapter 6.

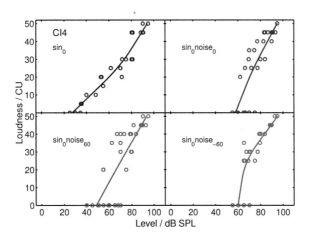

Figure 9.14: Individual results of CI user lkB in the Experiment in Chapter 6.

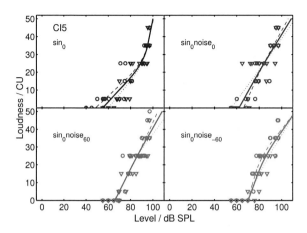

Figure 9.15: Individual results of CI user bs in the Experiment in Chapter 6.

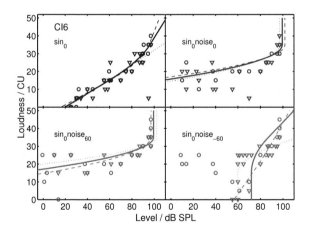

Figure 9.16: Individual results of CI user lkA in the Experiment in Chapter 6.

9.3 Individual results of the normal-hearing listeners in Chapter 7

Table 9.1: Individual results of the normal-hearing listeners (NH listeners) for the MAA task in Chapter 7 for center frequencies (cf) of 500 and 2500 Hz. The third column shows the results for the direct measurement of the minimum audible angle with the adaptive procedure. Columns 4 to 6 show the discrimination performance for angle differences of 5, 10 and 20° between reference and test signal measured with the constant stimulus procedure.

cf	NH listeners	MAA [°]	discrimination performance [%]		
			5°	10°	20°
500	NH1	2	92	100	96
	NH2	1.5	100	100	100
	NH3	2.5	100	100	96
	NH4	5	88	100	92
	NH5	6	64	84	100
	NH6	4	76	100	96
	NH7	3.5	100	96	96
	NH8	3	100	100	100
	NH9	2	96	100	100
	NH10	2.5	100	100	100
	NH11	3.5	88	96	100
2500	NH1	3	96	100	100
	NH2	3	100	100	100
	NH3	4	84	88	100

NH4	5	92	96	100
NH5	6.5	92	100	100
NH6	8.5	60	80	92
NH7	4.5	96	96	96
NH8	5.5	88	96	100
NH9	2.5	92	100	100
NH10	2.5	100	100	100
NH11	3.5	96	100	96

Table 9.2: Individual results of the normal-hearing listeners (NH listeners) for the MAMA task in Chapter 7 for velocities (vel) of 30 and $90°/s$. The third column shows the results for the direct measurement of the minimum audible movement angle with the adaptive procedure. Columns 4 and 5 show the discrimination performance for movement angles of 20 and $45°$ measured with the constant stimulus procedure.

vel	NH listeners	MAA [°]	discrimination performance [%]	
			$20°$	$45°$
30	NH1	10	100	92
	NH2	15	96	96
	NH3	15	88	100
	NH4	13.5	96	100
	NH5	11	80	96
	NH6	12	88	96
	NH7	10.5	96	100
	NH8	13	84	100
	NH9	12	96	100

	NH10	8.5	100	100
	NH11	9	100	100
90	NH1	15.5	100	92
	NH2	16	100	100
	NH3	21	76	100
	NH4	22	80	96
	NH5	19.5	80	96
	NH6	17	60	80
	NH7	20	92	100
	NH8	18.5	84	100
	NH9	11	96	96
	NH10	11.5	96	92
	NH11	11	80	96

Bibliography

Ahumada, A. J. and Lovell, J. (**1971**). "Stimulus features in signal detection," J. Acoust. Soc. Am. **49**, 1751–1756.

Agresti, A. (**1989**). "Tutorial on modeling ordered categorical response data," Psychol. Bull. **105**, 290–301.

Agresti, A. (**2002**). "Categorical data analysis," New York, NY: Wiley.

Alexander, J. M. and Lufti, R. A. (**2004**). "Informational masking in hearing-impaired and normal-hearing listeners: Sensation level and decision weights," J. Acoust. Soc. Am. **116**, 2234–2247.

Al-Salim, S. C., Kopun, J. G., Neely, S. T., Jesteadt, W., Steigemann, B., and Gorga, M. P. (**2010**). "Reliability of categorical loudness scaling and its relation to threshold," Ear Hearing **31**, 4, 567–578.

Altman, J. A. and Viskov, O. V. (**1977**). "Discrimination of perceived movement velocity used for fused auditory image in dichotic stimulation," J. Acoust. Soc. Am. **61**, 816–819.

Androulidakis, A. G., Jones, S. J. (**2006**). "Detection of signals in modulated and unmodulated noise observed using auditory evoked potentials." Clin. Neurophysiol. **117** (8), 1783–93.

ANSI S3.4 (**2007**). "Procedure for the computation of loudness of steady sounds," American National Standard S3.4-2007 (Standards Secreteriate, Acoustical Society of America).

Bibliography

Anweiler, A. K. and Verhey, J. L. (**2006**). "Spectral loudness summation for short and long signals as a function of level," J. Acoust. Soc. Am. **119**, 2919–2928.

Appell, J. E. (**2002**). "Loudness models for rehabilitative audiology," Ph.D. dissertation, Universität Oldenburg.

Bacon, E., Grimault, J. W., and Lee, J. H. (**2002**). "Spectral integration in bands of modulated or unmodulated noisea)," Hearing Res. **112** (1), 219–226.

Berg, B. G. and Robinson, D. E. (**1987**). "Multiple observations and internal noise," J. Acoust. Soc. Am. **81**, S33–S33.

Berg, B. G. (**1989**). "Analysis of weights in multiple observation tasks," J. Acoust. Soc. Am. **86**, 1743–1746.

Berg, B. G. (**1990**). "Observer efficiency and weights in a multiple observation task," J. Acoust. Soc. Am. **88**, 149–158.

Bland, J. M., and Altman, D. G. (**1995**). "Statistics notes: 12. Calculating correlation coefficients with repeated observations: Part 1 – correlation within subjects," Brit. Med. J. **310**, 446–446

Braida, L. D. and Durlach, N. I. (**1972**). "Intensity perception: II. Resolution in oneinterval paradigms," J. Acoust. Soc. Am. **51**, 4, 483–502.

Brand, T. and Hohmann, V. (**2002**). "An adaptive procedure for categorical loudness scaling," J. Acoust. Soc. Am. **112**, 4, 1597–1604.

Bremen, P. and Middlebrooks, J. C. (**2013**). "Weighting of Spatial and Spectro-Temporal Cues for Auditory Scene Analysis by Human Listeners," PLoS ONE **8**, 3, e59815.

Buss, E., Hall, J. W., and Grose, J. H. (**2012**). "CEffects of masker envelope irregularities on tone detection in narrowband and broadband noise maskers," Hearing Res. **294**, 73–81.

Buus, S., Zhang, L., and Florentine, M. (**1996**). "Stimulus-driven, time-varying weights for comodulation masking release," J. Acoust. Soc. Am. **99** (4), 2288–2297.

Buus, S., Florentine, M., and Poulsen, T. (**1997**). "Temporal integration of loudness, loudness discrimination, and the form of the loudness function," J. Acoust. Soc. Am. **101**, 669–680.

Calandruccio, L. and Doherty, K. A. (**2008**). "Spectral weighting strategies for hearingimpaired listeners measured using a correlational method," J. Acoust. Soc. Am. **123**, 2367–2378.

Cacace, A. T. and Margolis, R. H. (**1985**). "On the loudness of complex stimuli and its relationship to cochlear excitation.," J. Acoust. Soc. Am. **78**, 1568–1573.

Carlile, S. and Best, V. (**2002**). "Discrimination of sound source velocity in human listeners," J. Acoust. Soc. Am. **111** (2), 1026–1035.

Chalupper, J. (**2002**). "Perzeptive Folgen von Innenohrschwerhörigkeit: Modellierung, Simulation und Rehabilitation (Perceptive consequences of cochlear hearing impairment: modeling, simulations and rehabilitation)," Ph.D. thesis, Technische Universität München.

Chalupper, J. and Fastl, H. (**2002**). "Dynamic loudness model (DLM) for normal and hearing-impaired listeners," Acta Acust. united Ac. **88**, 378–386.

Chandler, D. W. and Grantham, D. W. (**1992**). "Minimum audible movement angle in the horizontal plane as a function of stimulus frequency and bandwidth, source azimuth, and velocity," J. Acoust. Soc. Am. **91**

(3), 1624–1636.

Dai, H. P., and Berg, B. G. (**1992**). "Spectral and temporal weights in spectral-shape discrimination," J. Acoust. Soc. Am. **92**, 1346–1355

Dau, T., Püschel, D.and Kollmeier, B. (**1996**). "A quantitative model of the effective signal processing in the auditory system. I. Model structure," J. Acoust. Soc. Am. **99**, 3615–3622.

DIN 45631 (**1991**). "Berechnung des Lautstärkepegels und der Lautheit aus dem Geräuschspektrum, Verfahren nach E. Zwicker (Procedure for calculating loudness level and loudness)," (Deutsches Institut für Normung e.V., Berlin, Germany).

DIN 45631/A1 (**2010**). "Berechnung des Lautstärkepegels und der Lautheit aus dem Geräuschspektrum, Verfahren nach E. Zwicker (Procedure for calculating loudness level and loudness - Zwicker method)," (Deutsches Institut für Normung e.V., Berlin, Germany).

Dittrich, K. and Oberfeld, D. (**2009**). "A comparison of the temporal weighting of annoyance and loudness," J. Acoust. Soc. Am. **126**, 3168–3178.

Doherty, K. A. and Lufti, R. A. (**1996**). "Spectral weights for overall level discrimination in listeners with sensorineural hearing loss," J. Acoust. Soc. Am. **99**, 1053–1058.

Dorfman, D. D. and Alf, E. (**1969**). "Maximum-likelihood estimation of parameters of Signal-Detection Theory and determination of confidence intervals - Ratingmethod data," J. Math. Psychol. **6**, 287–496.

Dorman, M. F. and Wilson, B. S. (**2004**). "The Design and Function of Cochlear Implants," American Scientist Online **92** (5), 436–444, http://www.americanscientist.org/issues/feature/the-design-and-function-of-cochlear-implants.

Ellermeier, W. and Schrödl, S. (**2000**). "Temporal weights in loudness summation," in *Bonnet C, editor. Fechner Day 2000 Proceedings of the 16th Annual Meeting of the International Society for Psychophysics*, Strasbourg: Université Louis Pasteur, 169–173.

Epp, B., Yasin, I., and Verhey, J. L., (**2013**). "Objective measures of binaural masking level differences and co-modulation masking release based on late auditory evoked potentials." Hearing Res. **306**, 21–28.

Epstein, M. and Florentine, M. (**2005**). "A test of the equal-loudness-ratio hypothesis using cross-modality matching functions," J. Acoust. Soc. Am. **118**, 907–913.

Epstein, M. (**2007**). "An introduction to induced loudness reduction," J. Acoust. Soc. Am. **122**, EL74–EL80.

Ernst, S. M. A., Uppenkamp, S., Verhey, J. L. (**2010**). "Cortical representation of release from auditory masking," Neuroimage **49**, 835–842.

European Parliament and Council of the European Union (**2002**). "Directive 2002/49/EC of the European Parliament and of the Council of 25 June 2002 relating to the assessment and management of environmental noise".

Fastl, H. and Zwicker, E. (**2007**). "Psychoacoustics – Facts and Models", 3rd edition, Springer Berlin Heidelberg, ISBN 978-3-540-68888-4.

Feddersen, W. E., Sandel, T. T., Teas, D. C., and Jeffress, L. A. (**1957**). "Localization of High-Frequency Tones," J. Acoust. Soc. Am. **29**, (9), 1633–1644.

Fletcher, H. and Steinberg, J. C. (**1924**). "The dependence of the loudness of a complex sound upon the energy in the various frequency regions of the sound," Phys. Rev. **24**, 306–317.

Fletcher, H. and Munson, W. A. (**1933**). "Loudness, its definition, measurement and calculation," J. Acoust. Soc. Am. **5**, 82–108.

Fletcher, H. and Munson, W. A. (**1937**). "Relation between loudness and masking," J. Acoust. Soc. Am. **9**, 1–10.

Florentine, M., Buus, S., and Poulsen, T. (**1996**). "Temporal integration of loudness as a function of level," J. Acoust. Soc. Am. **99**, 1633–1644.

Fowler, C. G., Mikami, C. M. (**1996**). "Phase effects on the middle and late auditory evoked potentials." J. Am. Acad. Audiol. **7** (1), 23–30.

Fruhmann, M., Chalupper, J., and Fastl, H. (**2003**). "Zum Einfluss von Innenohrschwerhörigkeit auf die Lautheitssummation (Influence of cochlear hearing impairment on spectral loudness summation)," in *DAGA 2003 - Fortschritte der Akustik, Proceedings of the 29th Annual Meeting of the Deutsche Gesellschaft für Akustik e.V.*, 253–254.

Gabriel, B., Kollmeier, B., and Mellert, V. (**1997**). "Influence of individual listener, measurement room and choice of test-tone levels on the shape of equal-loudness level contours," Acta Acust. united Ac. **83**, 670–683.

Gardner, M. B. and Gardner, R. S. (**1973**). "Problem of localization in the median plane: effect of pinnae cavity occlusion," J. Acoust. Soc. Am. **53** (2), 400–408.

Garnier, S., Micheyl, C., Berger-Vachon, C., and Colett, L. (**199**)). "Effect of signal duration on categorical loudness scaling in normal and in hearing-impaired listeners," Audiology **38**, (4), 196–201.

Glasberg, B. R. and Moore, B. C. J. (**2002**). "A model of loudness applicable to time-varying sounds," J. Audio Eng. Soc. **50**, 331–341.

Gleiss, N. and Zwicker, E. (**1964**). "Loudness function in the presence of masking noise," J. Acoust. Soc. Am. **36**, 393–394.

Gilkey, R. H. and Robinson, D. E. (**1986**). "Models of auditory masking: a molecular psychophysical approach," J. Audio Eng. Soc. **79**, 1499–1510.

Grantham, D. W. (**1986**). "Detection and discrimination of simulated motion of auditory targets in the horizontal plane", J. Acoust. Soc. Am. **79** (6), 1939–1949.

Grantham, S., Hornsby, B. W. Y., and Erpenbeck, E. A. (**2003**)). "Auditory spatial resolution in horizontal, vertical, and diagonal planes," J. Acoust. Soc. Am. **114**, (2), 1009–1022.

Grantham, S., Ashmead, C., Ricketts, C., Labadie, C., and Haynes, L. (**2007**)). "Horizontal-Plane Localization of Noise and Speech Signals by Postlingually Deafened Adults Fitted With Bilateral Cochlear Implants," Ear Hearing **28**, (4), 524–541.

Green, D. M. (**1958**). "Detection of multiple component signals in noise", J. Acoust. Soc. Am. **30**, 904–911.

Green, D. M. (**1964**). "Consistency of auditory detection judgments," Psychol. Rev. **71**, 392–407

Green, D. M., and Moses, D. M. (**1966**). "On the equivalence of two recognition measures of short-term memory," Psychol. Bull. **66**, 228–234

Green, D. M., and Swets, J. A. (**1966**). "Signal detection theory and psychophysics," New York: Wiley.

Grimm, G., Hohmann, C., and Verhey, J. L. (**2002**). "Loudness of fluctuating sounds," Acta Acust. united Ac. **88**, 359–368.

Guinan, J. J. (**1996**). "Physiology of Olivocochlear Efferents," in *The cochlea*, edited by P. Dallos, A. Popper, and R. Fay (Springer, New York), Vol. 8, 435–502.

Haeusler, R., Colburn, S., and Marr, E. (**1983**). "Sound Localization in Subjects with Impaired Hearing," Acta Otolaryngol. (Stockh), Printed in Sweden by Almqvist & Wiksell, Uppsala.

Hall, J. W., Haggard, M. P., and Fernandes, M. A. (**1984**). "Detection in noise by spectro-temporal pattern analysis," J. Acoust. Soc. Am. **76**, 50–56.

Hanley, J. A. (**1988**). "The robustness of the binormal assumptions used in fitting ROC curves," Med. Decis. Making **8**, 197–203

Harris, J. D. and Sergeant, R. L. (**1971**). "Monaural/binaural minimum audible angles for a moving sound source," J. Speech Hear. Res. **14**, 618–629.

Harris, D. M., and Dallos, P. (**1979**). "Forward masking of auditory-nerve fiber responses," J. Neurophysiol. **42**, 1083–1107.

Hartmann, W. M., and Rakerd, B. (**1989**). "On the minimum audible angle – A decision theory approach," J. Acoust. Soc. Am. **85** (5), 2031–2041.

Hawley, M. L., Litovsky, R. Y., and Colbourn, H. S. (**1999**). "Speech intelligibility and localization in a multi-source environment," J. Acoust. Soc. Am. **105**, 3436–3448.

Hawley, M. L., Litovsky, R. Y., and Culling, J. F. (**2004**). "The benefit of binaural hearing in a cocktail party: Effect of location and type of interferer," J. Acoust. Soc. Am. **115** (2), 833–843.

Heeren, W., Rennies, J., and Verhey, J. L. (**2011**). "Nicht-simultane spektrale Lautheitssummation (Non-simulataneous spectral loudness summation)," in *DAGA 2011 - Fortschritte der Akustik, Proceedings of the 37th Annual Meeting of the Deutsche Gesellschaft für Akustik e. V.*, 601–602.

Heeren, W., Rennies, J., and Verhey, J. L. (**2011**). "Spectral loudness summation of nonsimultaneous tone pulses," J. Acoust. Soc. Am. **130**, 3905–3915.

Heeren, W., Hohmann, V., Appell, J. E., and Verhey, J. L. (**2013**). "Relation between loudness in categorical units and loudness in phons and sones," J. Acoust. Soc. Am. **133** (4), EL314–EL319.

Hellbrück, J. and Moser, L. M. (**1985**). "Hörgeräte - Audiometrie: Ein computerunterstütztes psychologisches Verfahren zur Hörgeräteanpassung," Psycholog. Beiträge **27**, 494–508

Heller, O. (**1985**). "Hörfeldaudiometrie mit dem Verfahren der Kategorienunterteilung (KU)," Psycholog. Beiträge **27**, 478–493.

Henning, G. B. (**1974**). "Detectability of interaural delay in high-frequency complex waveforms," J. Acoust. Soc. Am. **55** (1) 84–90.

Hochberg, Y. (**1988**). "A sharper Bonferroni procedure for multiple tests of significance," Biometrika **75**, 800–802.

Hochmair, I., Nopp, P.,Jolly, C., Schmidt, M.,Schösser, H., Garnham, C., and Anderson, I. (**2006**). "MED-EL cochlear implants: state of the art and a glimpse into the future," Trends Amplif. **10**, 201–219.

Hohmann, Hohmann. and Kollmeier, B. (**1995**). "Weiterentwicklung und klinischer Einsatz der Hörfeldskalierun," Audiologische Akustik **34**, 48–59

Hosmer, D. W., and Lemeshow, S. (**2000**). "Applied logistic regression," New York: Wiley.

Hots, J. (**2014**). "Suprathreshold perception in normal-hearing and hearing impaired listeners", Ph.D. thesis, Logos Verlag Berlin GmbH, ISBN 978-3-8325-3758-6.

Hübner, R. and Ellermeier, W. (**1993**). "Additivity of loudness across critical bands: A critical test," Percept. Psychophys. **54**, 185–189

IEC 1970 (**1970**). "An IEC artificial ear, of the wide band type, for the calibration of earphones used in audiometry," (International Electrotechnical Commission, Geneva, Switzerland).

IEC 2009 (**2009**). "Reference zero for the calibration of audiometric equipment," (International Electrotechnical Commission, Geneva, Switzerland).

Ishida, I. M., Stapells, D. R. (**2004**). "Does the 40-Hz auditory steady-state response show the binaural masking level difference?" Ear Hearing **30** (6), 713–715.

ISO 1993 (**2003**). "Description, measurement and assessment of environmental noise," (International Organization for Standardization, Geneva, Switzerland).

ISO 226 (**2003**). "Acoustics – Normal equal-loudness-level contours," (International Organization for Standardization, Geneva, Switzerland).

ISO 389-8 (**2004**). "Acoustics – Reference zero for the calibration of audiometric equipment – SPart 8: Reference equivalent threshold sound pressure levels for pure tones and circumaural earphones," (International Organization for Standardization, Geneva, Switzerland).

ISO 16832 (**2006**). "Acoustics – Loudness scaling by means of categories," (International Organization for Standardization, Geneva, Switzerland).

ISO 532-1 (**2010**). "Acoustics – Method for calculating loudness level – Part 1: Stationary sounds," (International Organization for Standardization, Geneva, Switzerland).

Iverson, G., and Bamber, D. (**1997**). "The generalized area theorem in signal detection theory," in *Choice, decision, and measurement: essays in honor of R Duncan Luce, edited by R. D. Luce, and A. A. J Marley*. Mahwah, N.J.: L. Erlbaum, 301–318.

Jesteadt, W., Valente, D. L., and Joshi, S. (**2003**). "Perceptual weights for loudness judgments of 6-tone complexes," Poster presented at the Thirty-Fourth Annual Midwinter Research Meeting of the Association for Research in Otolaryngology. Baltimore.

Jonides, J. and Yantis, S. (**1988**). "Uniqueness of abrupt visual onset in capturing attention," Percept. Psychophys. **43**, 346–354.

Kiang, T., Watanabe, D., Thomas, D., and Clark, B. (**1965**). "Discharge Patterns of Single Fibers in the Cat's Auditory Nerve" The MIT Press, Cambridge, Massachusetts.

Kidd, G., Mason, C. R., Rohtla, T. L., and Deliwala, P.. (**1998**). "Release from masking due to spatial separation of sources in the identification of nonspeech auditory patterns," J. Acoust. Soc. Am. **104** (1), 422–431.

Keselman, H. J. (**1994**). "Stepwise and simultaneous multiple comparison procedures of repeated measures' means," J. Educ. Stat. **19**, 127–162.

Kohlrausch, A., Fassel, R., van der Heijden, M., Kortekaas, R., and van de Par, S. (**1997**). "Detection of tones in low-noise noise: Further evidence for the role of envelope fluctuations," Acustica **83**, 659–669.

Kortekaas, R., Buus, S., and Florentine, M. (**2003**). "Perceptual weights in auditory level discrimination," J. Acoust. Soc. Am. **113**, 3306–3322.

Leibold, L. J., Tan, H. Y., Khaddam, S., and Jesteadt, W. (**2007**). "Contributions of individual components to the overall loudness of a multitone complex" J. Acoust. Soc. Am. **121**, 2822–2831.

Leibold, L. J., Tan, H. Y., and Jesteadt, W. (**2009**). "Spectral weights for sample discrimination as a function of overall level," J. Acoust. Soc. Am. **125**, 339–346.

Levitt, H. (**1971**). "Transformed up-down methods in psychoacoustics," J. Acoust. Soc. Am. **49**, 467–477.

Litovsky, R. Y. (**1997**). "Developmental changes in the precedence effect: Estimates of minimum audible angle", J. Acoust. Soc. Am. **102** (3), 1739–1745.

Litovsky, R. Y. (**2005**). "Speech intelligibility and spatial release from masking in young children", J. Acoust. Soc. Am. **117**, 3091–3099.

Litovsky, R. Y., Parkinson, A., and Arcaroli, J. (**2009**). "Spatial hearing and speech intelligibility in bilateral cochlear implant users," Ear Hearing **30**, 419–431.

Litovsky, R. Y., Goupell, M. J., Godar, S., Grieco-Calub, T., Jones, G. L., Garadat, S. N., Agrawal, S., Kan, A., Todd, A., Hess, C., and Misurelli, J. (**2009**). "Studies on bilateral cochlear implants at the University of Wisconsin's Binaural Hearing and Speech Lab," J. Am. Acad. Audiol. **23** (6), 476–494.

Litovsky, R. Y. (**2012**). "Spatial release from masking", Acoustics Today **8** (6), 18–25.

Lochner, J. P. A. and Burger, J. F. (**1961**). "Form of the loudness function in the presence of masking noise," J. Acoust. Soc. Am. **33**, 1705–1707.

Loizou, P. C., Hu, Y., Litovsky, R. Y., Yu, G., Peters, R., Lake, J., and Roland, P. (**2009**). "Speech recognition by bilateral cochlear implant users in a cocktail-party setting," J. Acoust. Soc. Am. **125**, 372–383.

Long, C. J., Carlyon, R. P., Litovsky, R. Y., and Downs, D. H. (**2006**). "Binaural Unmasking with Bilateral Cochlear Implants," JARO **7**, 352–360.

Lu, T., Litovsky, R. Y., and Zeng, F.-G. (**2010**). "Binaural masking level differences in actual and simulated bilateral cochlear implant listeners," J. Acoust. Soc. Am. **127 3**, 1479–1490.

Lufti, R. A. (**1989**). "Informational processing of complex sound. I: Intensity discrimination," J. Acoust. Soc. Am. **86**, 934–944.

Lufti, R. H. and Wang, W. (**1999**). "Correlational analysis of acoustic cues for the discrimination of auditory motion," J. Acoust. Soc. Am. **106** (2), 919–928.

Lufti, R. A., and Jesteadt, W. (**2006**). "Molecular analysis of the effect of relative tone level on multitone pattern discrimination," J. Acoust. Soc. Am. **120**, 3853–3860.

Macmillan, N. A.., Rotello, C. M., and Miller, J. O. (**2004**). "The sampling distributions of Gaussian ROC statistics," Percept. Psychophys. **66**, 406–421.

Macmillan, N. A.., and Creelman, C. D. (**2005**). "Detection theory: A user's guide," Mahwah, NJ: Lawrence Erlbaum Associates.

Mauermann, M., Long, G. R., and Kollmeier, B. (**2004**). "Fine structure of hearing threshold and loudness perception," J. Acoust. Soc. Am. **116**, 1066–1080.

McCullagh, P. (**1980**). "Regression models for ordinal data," J. R. Statist. Soc. B. **42**, 109–142.

McFadden, D. (**1968**). "Masking-Level Differences Determined with and without Interaural Disparities in Masker Intensity," J. Acoust. Soc. Am.

44 (1), 212–223.

McFarland, D. J., and Cacace, A. T. (**1992**). "Aspects of short-term acoustic recognition memory: Modality and serial position effects," Audiology **31**, 342–352.

Metz, C. E., Herman, B. A., and Shen, J. H. (**1998**). "Maximum likelihood estimation of receiver operating characteristic (ROC) curves from continuously-distributed data," Stat. Med. **17**, 1033–1053.

Middlebrooks, J. C., Makous, J. C., and Green, D. M. (**1989**). "Directional sensitivity of sound-pressure levels in the human ear canal," J. Acoust. Soc. Am. **86**, 89–108.

Middlebrooks, J. C., and Green, D. M. (**1991**). "Sound localization by human listeners," Annu. Rev. Psychol. **42**, 135–159.

Mills, A. W. (**1958**). "On the Minimum Audible Angle," J. Acoust. Soc. Am. **30**, 4, 237–246.

Mills, A. W. (**1960**). "Lateralization of high-frequency tones," J. Acoust. Soc. Am. **32**, 132–134.

Mondor, T. A., and Morin, S. R. (**2004**). "Primacy, recency, and suffix effects in auditory short-term memory for pure tones: Evidence from a probe recognition paradigm," Can. J. Exp. Psychol. - Rev. Can. Psychol. Exp. **58**, 206–219.

Moore, B. C. J. and Shailer, M. J. (**1990**). "Comodulation masking release as a function of level," J. Acoust. Soc. Am. **90** (2), 829–835.

Moore, B. C. J. (**2003**). *An Introduction to the Psychology of Hearing*, 5th edition (Elsevier Academic Press, San Diego, USA), 107 ff.

Moore, B. C. J. and Glasberg, B. R. (**1996**). "A revision of Zwicker's loudness model," Acta Acust. united Ac. **82**, 335–345.

Moore, B. C. J., Glasberg, B. R., and Baer, T. (**1997**). "A model for the prediction of thresholds, loudness and partial loudness," J. Audio Eng. Soc. **45**, 224–239.

Munson, W. A. (**1947**). "The growth of auditory sensation," J. Acoust. Soc. Am. **19**, 584–591.

Nava, E., Bottari, D., Bonfioli, F., Beltrame, M. A., and Pavani, F. (**2009**). "Spatial hearing with a single cochlear implant in late-implanted adults," Hearing Res. **255**, 91–98.

Nieder, B., Florentine, M., Buus, S., and Scharf, B. (**2003**). "Interactions between test- and inducer-tone durations in induced loudness reduction," J. Acoust. Soc. Am. **114**, 2846–2855.

Niese, H. (**1959**). "Die Trägheit der Lautstärkebildung in Abhängigkeit vom Schallpegel," Hochfrequenztechn. u. Elektroak. **68**, 143.

Niparko, J. K., Kirk, K. I., Mellon, N. K., Robbins, A. M., Tucci, D.L., and Wilson, D. S. (Eds.) (**2000**). "Cochlear Implants: Principles & Practices", Lippincott Williams & Wilkins, Philadelphia, ISBN 0-7817-1782-5.

Nitschmann, M., Verhey, J. L., and Kollmeier, B. (**2009**). "The role of across-frequency processes in dichotic listening conditions," J. Acoust. Soc. Am. **126** (6), 3188–3198.

Nopp, P., Schleich, P., and D'Haese, P. (**2004**). "Sound Localization in Bilateral Users of MED-EL COMBI 40/40+ Cochlear Implants", Ear Hearing **25** (3), 205–214.

Bibliography

Oberfeld, D. (**2008**). "Does a rhythmic context have an effect on perceptual weights in auditory intensity processing?," Can. J. Exp. Psychol.-Rev. Can. Psychol. Exp. **62**, 24–32.

Oberfeld, W. (**2009**). "The temporal weighting of the loudness of time-varying sounds reflects both sensory and cognitive processes," in *Fechner Day 2009 Proceedings of the 25th Annual Meeting of the International Society for Psychophysics, edited by M. A. Elliott, S. Antonievic, S. M. Berthaud, B. Bargary, and C. Martyn.* Galway, Ireland: International Society for Psychophysics, 261–266.

Oberfeld, D. (**2010**). "Electrophysiological correlates of intensity resolution under forward masking," in *The Neurophysiological Bases of Auditory Perception, edited by E. A. Lopez-Poveda, A. R. Palmer, and R. Meddis (Springer, New York),* 99–110.

Oberfeld, D. and Plank, T. (**2011**). "The temporal weighting of loudness: effects of the level profile," Atten. Percept. Psychophys. **73**, 189–208.

Oberfeld, D. and Franke, T. (**2012**). "Evaluating the robustness of repeated measures analyses: The case of small sample sizes and non-normal data," Behav. Res. Meth. **http:/dx.doi.org/10.3758/s13428-012-0281-2**.

Oberfeld, D., Heeren, W., Rennies, J. and Verhey, J. L. (**2012**). "Spectro-Temporal Weighting of Loudness," PLoS ONE **7** (11), e50184.

Ogura, Y., Suzuki, Y., and Sone, T. (**1991**). "A temporal integration model for loudness perception of repeated impulsive sounds," J. Acoust. Soc. Jpn. **12**, 1–11.

Patterson, R. D. (**1976**). "auditory filter shapes derived with noise stimuli," J. Acoust. Soc. Am. **59** (3), 640–654.

Pavlov, I. P. (**1927**). "Conditioned reflexes: An investigation of the physiological activity of the cerebral cortex,", translated and edited by

C. V. Anrep, (London: Oxford University Press.)

Pedersen, B. (**2006**). "Auditory Temporal Resolution and Integration. Stages of Analyzing Time-Varying Sounds," Ph.D. thesis, Aalborg University.

Pedersen, B. and Ellermeier, W. (**2008**). "Temporal weights in level discrimination of time-varying sounds," J. Acoust. Soc. Am. **123**, 963–972.

Perrott, D. R. and Musicant, A. D. (**1977**). "Minimum auditory movement angle: Binaural localization of moving sound sources," J. Acoust. Soc. Am. **62** (6), 1463–1466.

Perrott, D. R. and Tucker, J. (**1988**). "Minimum audible movement angle as a function of signal frequency and the velocity of the source," J. Acoust. Soc. Am. **83** (4), 1522–1527.

Perrott, D. R. and Marlborough, K. (**1989**). "Minimum audible movement angle: Marking the end points of the path traveled by a moving sound source," J. Acoust. Soc. Am. **85** (4), 1773–1775.

Perrott, D. R. and Pacheco, S. (**1989**). "Minimum audible angle thresholds for broadband noise as a function of the delay between the onset of the lead and lag signals," J. Acoust. Soc. Am. **85** (6), 2669–2672.

Perrott, D. R., Marlborough, K., and Merrill, P. (**1989**). "Minimum audible angle thresholds obtained under conditions in which the precedence effect is assumed to operate," J. Acoust. Soc. Am. **85** (1), 282–288.

Perrott, D. R. and Saberi, K. (**1990**). "Minimum audible angle thresholds for sources varying in both elevation and azimuth," J. Acoust. Soc. Am. **87** (4), 1728–1731

Plack, C. J. and Moore, B. C. J. (**1990**). "Temporal window shape as a function of frequency and level," J. Acoust. Soc. Am. **87**, 2178–2187.

Bibliography

Plank, T., and Ellermeier, W. (**2004**). "Discriminating temporal loudness patterns in the absence of overall level cues," in *Société Francaise d'Acoustique, Deutsche Gesellschaft für Akustik, editors. Proceedings of the Joint Congress CFA/DAGA '04*. Strasbourg, 397–398.

Plank, T. (**2005**). "Auditive Unterscheidung von zeitlichen Lautheitsprofilen)," Ph.D. thesis, Universität Regensburg..

Plenge, G. (**1974**). "On the differences between localization and lateralization," J. Acoust. Soc. Am. **56** (3), 944–951.

Plomp, R., and Mimpen, A. M. (**1981**). "Effect of the orientation of the speaker's head and the azimuth of a noise source on the speech reception threshold for sentences," Acustica **48**, 325–328.

Port, E. (**1963**a). "Über die Lautstärke kurzer Schallimpulse (Loudness of short sound pulses)," Acustica **13**, 212–223.

Port, E. (**1963**b). "Zur Lautstärke und Lautstärkemessung von pulsierenden Geräuschen (Loudness and loudness measurement of pulsating sounds)," Acustica **13**, 224–233.

Postman, L., and Phillips, L. W. (**1965**). "Short-term temporal changes in free recall," J. Exp. Psychol. **17**, 132–138.

Poulsen, T. (**1981**). "Loudness of tone pulses in a free field," J. Acoust. Soc. Am. **69**, 1786–1790.

Rasmussen, A. N., Olsen, S.O., Borgkvist, B. V., Nielsen, L. H. (**1998**). "A Long-term Test-Retest Reliability of Category Loudness Scaling in Normal-Hearing Subjects Using Pure-tone Stimuli." Scand. Audiol. **27**, 161–167.

Rayleigh, Lord (**1907**). "On our perception of sound direction," Philos. Mag. **13**, 214–232.

Rennies, J. (**2008**). "Perception and modeling of loudness of time-varying sounds," Master thesis, Engineering Physics, Carl von Ossietzky University Oldenburg.

Rennies, J. and Verhey, J. (**2009**). "Temporal weighting in loudness of broadband and narrowband signals (L)," J. Acoust. Soc. Am. **126**, 951–954.

Rennies, J., Verhey, J. L., Chalupper, J., and Fastl, H. (**2009**). "Modeling temporal effects of spectral loudness summation," Acta Acust. united Ac. **95**, 1112–1122.

Rennies, J., Verhey, J. L., and Fastl, H. (**2010**). "Comparison of loudness models for time-varying sounds," Acta Acust. united Ac. **96**, 383–396.

Ross, B., Fujioka, T., Tremblay, K. L., Picton, T. W. (**2007**). "Aging in binaural hearing begins in mid-life: Evidence from cortical auditory-evoked responses to changes in interaural phase." J. Neurosci. **27** (42), 11172–11178.

Rupp, A., Las, L., Nelken, I. (**2007**). "Neuromagnetic representation of comodulation masking release in the human auditory cortex". In: Kollmeier, B and Hohmann, V and Mauermann, M and Verhey, J and Klump, G and Langemann, U and Uppenkamp, S. (Ed.), *Hearing – From Sensory Processing to Perception.*, Springer-Verlag Berlin, Germany, 125–132.

Saberi, K. and Perrott, D. R. (**1990**). "Minimum audible movement angles as a function of sound source trajectory," J. Acoust. Soc. Am. **88** (6), 2639–2644.

Saberi, K., Dostal, L., Sadralodaba, T., Bull, V., and Perrott, D. R. (**1991**). "Free-field release from masking," J. Acoust. Soc. Am. **90** 3, 1355–1370.

Sasaki, T., Kawase, T., Nakasato, N., Kanno, A., Ogura, M., Tominaga, T. and Kobayashi, T. (**2005**). "Neuromagnetic evaluation of binaural unmasking," Neuroimage **25**, 684–689.

Scharf, B. (**1962**). "Loudness of complex sounds as a function of the number of components", J. Acoust. Soc. Am. **31**, 783–785.

Scharf, B. (**1962**). "Loudness summation and spectrum shape", J. Acoust. Soc. Am. **34**, 228–233.

Scharf, B. (**1970**). *Frequency analysis and periodicity detection in hearing*, Chapter: Loudness and frequency selectivity at short duration (Leiden: Sijthoff), 455–462.

Schleich, P., Nopp, P., and D'Haese, P. (**2004**). "Head shadow, squelch, and summation effects in bilateral users of the MED-EL COMBI 40/40+ cochlear implant", Ear Hearing **25** (3), 197–204.

Schneider, B. (**1988**). "The additivity of loudness across critical bands: A conjointmeasurement approach", Percept. Psychophys. **43**, 211–222.

Schoen, F., Mueller, J., Helms, J., and Nopp, P. (**2005**). "Sound Localization and Sensitivity to Interaural Cues in Bilateral Users of the Med-El Combi 40/40+Cochlear Implant System," Otol. Neurotol. **26** (3), 429–437.

Sechenov, I. M., (**1965**). "Reflexes of the brain (original publication 1863)," Cambridge, Mass.: M.I.T. Press..

Seeber, B. U. and Fastl, H. (**2003**). "Subjective selection of non-individual head-related transfer functions," in *Proceedings of the 2003 International Conference on Auditory Display*, Boston, MA, USA.

Seeber, B. U. and Fastl, H. (**2008**). "Localization cues with bilateral cochlear implants,", J. Acoust. Soc. Am. **123** 2, 1030–1042.

Senn, P., Kompis, M., Vischer, M., and Haeusler, R.,(**2005**). "AMinimum Audible Angle, Just Noticeable Interaural Differences and Speech Intelligibility with Bilateral Cochlear Implants Using Clinical Speech Processors", Audiol. Neurotol., **40**, 342–352.

Smith, R. L. (**1977**). "Short-term adaptation in single auditory nerve fibers: some poststimulatory effects," J. Neurophysiol. **40**, 10198–1112.

Soderquist, D. R. and Schilling, R. D. (**1990**). "Loudness and the binaural masking level difference," Bull. Psychonomic. Soc. **28**, 553–555.

Sorkin, R. D., Sorkin, R. D., and Sorkin, R. D.,(**1987**). "A detection theory method for the analysis of auditory and visual displays", Proceedings of the 31st Annual Meeting of the Human Factors Society, 1184–1184.

Springer Handbook of Acoustics(**2007**). Rossing, T. D. (Editor), Springer Science+Business Media, LLC New York, ISBN 978-0-387-30446-5, page 435 .

Stecker, G. C., and Hafter, E. R. (**2002**). "Temporal weighting in sound localization," J. Acoust. Soc. Am. **112**, 1046–1057.

Stecker, G. C., and Brown, A. D. (**2010**). "Temporal weighting of binaural cues revealed by detection of dynamic interaural differences in high-rate Gabor click trains," J. Acoust. Soc. Am. **127**, 3092–3103.

Stevens, S. S. (**1956**). "The direct estimation of sensory magnitudes - Loudness," Am. J. Psychol. **69**, 1–25.

Stevens, S. S. (**1957**). "On the psychophysical law," Psychol. Rev. **64**, 153–181.

Strybel, T. Z. and Fujimoto, K. (**2000**). "Minimum audible angles in the horizontal and vertical planes: Effects of stimulus onset asynchrony and burst duration," J. Acoust. Soc. Am. **108** (6), 3092–3095

Suzuki, Y., and Takeshima, H. (**2004**). "Equal-loudness-level contours for pure tones," J. Acoust. Soc. Am. **116**, 918–933.

Swets, J. A. (**1986**). "Indices of discrimination or diagnostic accuracy: their ROCs and implied models", Psychol. Bull. **99**, 100–117.

Steinberg, T. H. and Gardner, D. P. (**1972**). "Suprathreshold Binaural Unmasking," J. Acoust. Soc. Am. **51**, 621–624.

Trahiotis, C., and Stern, R. M. (**1989**). "TLateralization of bands of noise: effects of bandwidth and differences of interaural time and intensity," J. Acoust. Soc. Am. **86**, 1285–1293.

Turner, M. D., and Berg, B. G. (**2007**). "Temporal limits of level dominance in a sample discrimination task," J. Acoust. Soc. Am. **120**, 1848–1851.

Ulanovsky, N., Las, L., and Nelken, I. (**2003**). "Processing of low-probability sounds by cortical neurons,", Nat. Neurosci. **6**, 391–398.

Ulrich, R., and Vorberg, D. (**2009**). "Estimating the difference limen in 2AFC tasks: Pitfalls and improved estimators,", Atten. Percept. Psychophys. **71**, 1219–1227.

van Beurden, M. F. and Dreschler, W. A. (**2007**). "Duration dependency of spectral loudness summation, measured with three different experimental procedures,", in *Hearing - From Sensory Processing to Perception*, edited by B. Kollmeier, G. Klump, V. Hohmann, U. Langemann, M. Mauermann, S. Uppenkamp, and J.-L. Verhey, (Springer, Berlin - Heidelberg), 237–246 .

van de Par, S., and Kohlrausch, A. (**1997**). "A new approach to comparing binaural masking level differences at low and high frequencies," J. Acoust. Soc. Am. **101** (3), 1671–1680.

van Hoesel, R. J. M., Ramsden, R., and O'Driscoll, M. (**2002**). "Sound-Direction Identification, Interaural Time Delay Discrimination, and Speech Intelligibility Advantages in Noise for a Bilateral Cochlear Implant User," Ear Hearing **23** (2), 137–149.

van Hoesel, R. J. M., and Tyler, R. S. (**2003**). "Speech perception, localization, and lateralization with bilateral cochlear implants," J. Acoust. Soc. Am. **113** (3), 1617–1630.

van Hoesel, R. J. M., Böhm, M., Pesch, J., Vandali, A., Battmer, R. D, and Lenarz, T. (**2008**). "Binaural speech unmasking and localization in noise with bilateral cochlear implants using envelope and fine-timing based strategies," J. Acoust. Soc. Am. **123** (4), 2249–2263.

van Wanrooij, M. M., and van Opstal, A. J. (**2004**). "Contribution of head shadow and pinna cues to chronic monaural sound localisation," J. of Neurosci. **24**, 4163–4171.

Verhey, J. L. (**1999**). "Psychoacoustiscs of Spectro-temporal effects in masking and loudness perception," (BIS - Verlag Oldenburg), ISBN 3-8142-0662-2, http://docserver.bis.uni-oldenburg.de/publikationen/bisverlag/verpsy99/verpsy99.html (date last viewed 8/2/11).

Verhey, J. L., Dau, T., and Kollmeier, B. (**1999**). "Within-channel cues in comodulation masking release (CMR): Experiments and model predictions using a modulation-filterbank model," J. Acoust. Soc. Am. **106**, 2733–2745.

Verhey, J. L. and Kollmeier, B. (**2002**). "Spectral loudness summation as a function of duration," J. Acoust. Soc. Am. **111**, 1349–1358.

Verhey, J. L., Pressnitzer, D., and Winter, I.M. (**2003**). "The psychophysics and physiology of comodulation masking release," Exp. Brain Res. **153**, 405–417.

Verhey, J. L. and Uhlemann, M. (**2008**). "Spectral loudness summation for sequences of short noise bursts," J. Acoust. Soc. Am. **123**, 925–934.

Verhey, J. L. and Ernst, S. M. A. (**2009**). "Comodulation masking release for regular and irregular modulators," Hearing Res. **253**, 97–106.

Verhey, J. L., Heeren, W., and Rennies, J. (**2011**). "Non-simultaneous across-frequency interaction in loudness (A)," in *Assoc. Res. Otolaryngol. Abstract Book*, volume 34, 158.

Verhey, J. L. and Heise, S. J. (**2012**). "Suprathreshold perception of tonal components in noise under conditions of masking release," Acta Acust. united Ac. **153**, 451–460.

Verschuur, C. A., Lutman, M. E., Ramsden, R., Greenham, P., and O'Driscoll, M. (**2005**). "Auditory Localization Abilities in Bilateral Cochlear Implant Recipients," Otol. Neurotol. **26** (5), 965–971.

Wack, D. S., Cox, J. L., Schirda, C. V., Magnano, C. R., Sussman, J. E., Henderson, D., Burkard, R. F. (**2012**). "Functional Anatomy of the Masking Level Difference, an fMRI Study", PLoS ONE **7** (7), e41263.

Wenzel, E. M., Arruda, M., Kistler, D. J., and Wightman, F. L. (**1993**). "Localization using nonindividualized head-related transfer functions," J. Acoust. Soc. Am. **94** (1), 111–123.

Wightman, F. L. and Kistler, D. J. (**1989a**). "Headphone simulation of free-field listening I: Stimulus synthesis," J. Acoust. Soc. Am. **85**, 858–867.

Wightman, F. L. and Kistler, D. J. (**1989b**). "Headphone simulation of free-field listening II: Psychophysical validation," J. Acoust. Soc. Am. **85**, 868–878.

Wilson, B. S., Finley, C. C., Lawson, D. T., Wolford, R. D., Eddington, D. K., and Rabinowitz, W. M. (**1991**). "Better speech recognition with cochlear implants," Nature **352**, 236–238.

Wilson, B. S. and Dorman, F. M. (**2008**). "Cochlear implants: Current designs and future possibilities," J. Rehabil. Res. Dev. **45** (5), 695–730.

Wong, W. Y. S., Stapells, D. R. (**2004**). "Brain stem and cortical mechanisms underlying the binaural masking level difference in humans: an auditory steady-state response study." Ear Hearing **5** (1), 57–67.

World Health Organization (**1999**). "Guide for community noise".

Yeshurun, Y., Carrasco, M., and Maloney, L. T. (**2008**). "Bias and sensitivity in two-interval forced choice procedures: Tests of the difference model," Vision. Res. **48**, 1837–1851.

Yost, W. A., Wightman, F. L., and Green, D. M. (**1971**). "Lateralization o f filtered clicks," J. Acoust. Soc. Am. **50**, 1526–1531.

Zeng, F.-G. and Turner, C. W. (**1992**). "Intensity discrimination in forward masking," J. Acoust. Soc. Am. **92**, 782–787.

Zwicker, E. and Feldtkeller, R. (**1955**). ""Über die Lautst¨arke von gleichf¨ormigen Geräuschen (Loudness of uniform sounds)," Acoustica **5**, 303–316.

Zwicker, E., Flottorp, G., and Stevens, S. S. (**1957**). "Critical band width in loudness summation," J. Acoust. Soc. Am. **29**, 548–557.

Zwicker, E. (**1958**). "Über psychologische und methodische Grundlagen der Lautheit (On psychological and methodical principles of loudness)," Acustica **8**, 237–258.

Zwicker, E. and Scharf, B. (**1965**). "A model of loudness summation," Psychol. Rev. **72**, 3–26.

Zwicker, E. (**1969**). "Der Einfluss der zeitlichen Struktur von Tönen auf die Addition von Teillautheiten (Influence of the temporal structure of tones on the summation of partial loudnesses)," Acustica **21**, 16–25.

Zwicker, E. (**1974**). "Loudness and excitation patterns of strongly frequency modulated tones," in *Sensation and measurement, papers in honor of S.S. Stevens*, edited by H. Moskowitz, B. Scharf, and J. Stevens (D. Reidel, Dordrecht - Boston), 325–335.

Zwicker, E. (**1977**). "Procedure of calculating loudness of temporally variable sounds," J. Acoust. Soc. Am. **62**, 675–682.

Zwicker, E. (**1984**). "Dependence of post-masking on masker duration and its relation to temporal effects in loudness," J. Acoust. Soc. Am. **57**, 219–223.

Zwicker, E. and Henning, G. B. (**1991**). "On the effect of interaural phase differences on loudness," Hearing Res. **53**, 141–152.

Zwicker, E. and Fastl, H. (**1999**). "Psychoacoustics – Facts and Models", 2rd edition, Springer Berlin Heidelberg, ISBN 978-3-540-65063-8.

Zwislocki, J. J. (**1969**). "Temporal summation of loudness: An analysis," J. Acoust. Soc. Am. **46**, 431–441.

Zwislocki, J. J., Ketkar, I., Cannon, M. W., and Nodar, R. H.(**1974**). "Loudness enhancement and summation in pairs or short sound bursts," Percept. Psychophys. **16**, 91–95.

List of Figures

List of Tables

Danksagungen

Besonders herzlich möchte ich Prof. Dr. Jesko L. Verhey für die umfangreiche Betreuung meiner Arbeit danken. Während der gesamten Zeit meiner Arbeit konnte ich mich auf seine Hilfe stets verlassen.

Auch der Abteilung für Experimentelle Audiologie möchte ich danken, die mir immer Rat und Tat zur Seite stand. Insbesondere gilt mein Dank hier zum einen Katrin Mentzel, die mir eine große Hilfe bei der Durchführung der in Kapitel 5 beschriebenen Experimente war. Zum Anderen bin ich Michael Ziese sehr dankbar dafür, dass er geduldig all sein Wissen um und über Cochlear-Implantate und deren Anpassung mit mir geteilt hat.

Ein großes Dankeschön geht an Jan Rennies, Daniel Oberfeld, Volker Hohmann und Jens E. Appell für die effektive und unkomplizierte Zusammenarbeit. Des Weiteren möchte ich Brian C. J. Moore und einigen anonymen Reviewern für ihre kritische Auseinandersetzung mit den eingereichten Manuskripten der veröffentlichten Teile dieser Arbeit danken.

Neben den Dank für all die wissenschaftlichen Unterstützung geht mein Dank auch an alle Studenten, Patienten und anderen Freiwilligen, die immer wieder bereit waren an meinen Studien teilzunehmen, sowie an Christin und Thomas ohne die ich gar nicht die Berechtigung zur Promotion erteilt bekommen hätte.

Zu guter Letzt gilt mein Dank natürlich all den Menschen, insbesondere meiner Familie und meinen Freunden, die mich durch die Zeit meiner Promotion begleitet und mich in jeder Situation unterstützt und vorangetrieben haben.